Birkhäuser

Advanced Courses in Mathematics
CRM Barcelona

Centre de Recerca Matemàtica

Managing Editor:
Carles Casacuberta

More information about this series at http://www.springer.com/series/5038

Feng Dai • Yuan Xu

Analysis on h-Harmonics and Dunkl Transforms

Editor for this volume:
Sergey Tikhonov, ICREA and CRM, Barcelona

Birkhäuser

Feng Dai
Department of Mathematics
and Statistical Sciences
University of Alberta
Edmonton, AB, Canada

Yuan Xu
Department of Mathematics
University of Oregon
Eugene, OR, USA

ISSN 2297-0304 ISSN 2297-0312 (electronic)
Advanced Courses in Mathematics - CRM Barcelona
ISBN 978-3-0348-0886-6 ISBN 978-3-0348-0887-3 (eBook)
DOI 10.1007/978-3-0348-0887-3

Library of Congress Control Number: 2014959869

Mathematics Subject Classification (2010): Primary: 41A10, 42B15; Secondary: 42B25, 42B08, 41A17

Springer Basel Heidelberg New York Dordrecht London

Printed on acid-free paper

Springer Basel AG is part of Springer Science+Business Media (www.birkhauser-science.com)

Contents

Preface

These lecture notes were written as an introduction to Dunkl harmonics and Dunkl transforms, which are extensions of ordinary spherical harmonics and Fourier transforms with the usual Lebesgue measure replaced by weighted measures.

The theory was initiated by C. Dunkl and subsequently developed by many authors in the past two decades. In this theory, the role of orthogonal groups, which provide the underline structure for the ordinary Fourier analysis, is played by a finite reflection group, the partial derivatives are replaced by the Dunkl operators, which are a family of commuting first order differential and difference operators, and the Lebesgue measure is replaced by a weighted measure with the weight function h_κ invariant under the reflection group, where κ is a parameter. The theory has a rich structure parallel to that of Fourier analysis, which allows us to extend many classical results to the weighted setting, especially in the case of h-harmonics, which are the analogues of ordinary spherical harmonics. There are still many problems to be solved and the theory is still at its infancy, especially in the case of Dunkl transform. Our goal is to give an introduction to what has been developed so far.

The present notes were written for people working in analysis. Prerequisites on reflection groups are kept to a bare minimum. In fact, even assuming the group is $\mathbb{Z}_2^d$, which requires essentially no prior knowledge of reflection groups, a reader can still gain access to the essence of the theory and to many highly non-trivial results, where the weight function h_κ is simply

$$h_\kappa(x) = \prod_{i=1}^{d} |x_i|^{\kappa_i}, \qquad \kappa_i \ge 0, \ \ 1 \le i \le d,$$

the surface measure $d\sigma$ on the sphere $\mathbb{S}^{d-1}$ is replaced by $h_\kappa^2 d\sigma$, and the Lebesgue measure dx on $\mathbb{R}^d$ is replaced by $h_\kappa^2 dx$.

To motivate the weighted results, we give a brief recount of basics of ordinary spherical harmonics and the Fourier transform in the first chapter, which can be skipped altogether. The Dunkl operators and the intertwining operator between partial derivatives and the Dunkl operators, are introduced and discussed in the second chapter. The intertwining operator plays a key role in the theory as it appears in the concise formula for the reproducing kernel of the h-spherical harmonics and in the definition of the Dunkl transform. The next three chapters are devoted to analysis on the sphere. The third chapter is an introduction to h-harmonics and essential results on harmonic analysis in the weighted space. The Littlewood–Paley theory on the sphere is developed in the fourth chapter, and is used to establish a Marcinkiewicz type multiplier theorem on the weighted sphere. As an application, two inequalities, the sharp Jackson and sharp Marchaud inequalities, are established in the fifth chapter, which are useful for approximation theory and in the embedding theory of function spaces. The final two chapters are devoted to the Dunkl transform. The sixth chapter is an introduction to Dunkl transforms, where the basic results are developed in detail. The Littlewood–Paley theory and a multiplier theorem are established in the seventh chapter, using a transference between h-harmonic expansions on the sphere and the Dunkl transform in $\mathbb{R}^d$.

The topics reflect the authors' choice. There are many results for Dunkl transforms on the real line (where the measure is $|x|^\kappa dx$) that we did not discuss, since the setting on the real line is closely related to the Hankel transforms and often cannot even be extended to the $\mathbb{Z}_2^d$ case in $\mathbb{R}^d$. There are also results on partial differential-difference equations, in analogy to PDE, that we did not discuss. Because of the explicit formula for the intertwining operator, the case $\mathbb{Z}_2^d$ has seen far more, and deeper, results, especially in the case of analysis on the sphere such as those for Cesàro means. We chose the Littlewood–Paley theory and the multiplier theorem, as this part is relatively complete and the results are related in the two settings, the sphere and the Euclidean space.

These lecture notes were written for the advanced courses in the program *Approximation Theory and Fourier Analysis* at the Centre de Recerca Matemàtica, Barcelona. We are grateful to the CRM for the warm hospitality during our two months stay, to the participants in our lectures, and thank especially the organizer of the program, Sergey Tikhonov from CRM, for his great help. We gratefully acknowledge the support received from NSERC Canada under grant RGPIN 311678-2010 (F.D.), from National Science Foundation under grant DMS-1106113 (Y.X.), and from the Simons Foundation (# 209057 to Y. X.).

Edmonton, Alberta, and Eugene, Oregon
September, 2014

Feng Dai
Yuan Xu

Chapter 1

Introduction: Spherical Harmonics and Fourier Transform

The purpose of these lecture notes is to provide an introduction to two related topics: h-harmonics and the Dunkl transform. These are extensions of the classical spherical harmonics and the Fourier transform, in which the underlying rotation group is replaced by a finite reflection group. This chapter serves as an introduction, in which we briefly recall classical results on the spherical harmonics and the Fourier transform. Since all results are classical, no proof will be given.

1.1 Spherical harmonics

First we introduce several notations that will be used throughout these lecture notes.

For $x \in \mathbb{R}^d$, we write $x = (x_1, \ldots, x_d)$. The inner product of $x, y \in \mathbb{R}^d$ is denoted by $\langle x, y \rangle := \sum_{i=1}^d x_i y_i$ and the norm of x is denoted by $\|x\| := \sqrt{\langle x, x \rangle}$. Let $\mathbb{S}^{d-1} := \{x \in \mathbb{R}^d : \|x\| = 1\}$ denote the unit sphere of $\mathbb{R}^d$, and let $\mathbb{N}_0$ denote the set of nonnegative integers. For $\alpha = (\alpha_1, \ldots, \alpha_d) \in \mathbb{N}_0^d$, a monomial x^α is a product $x^\alpha = x_1^{\alpha_1} \cdots x_d^{\alpha_d}$, which has degree $|\alpha| := \alpha_1 + \cdots + \alpha_d$.

A homogeneous polynomial P of degree n is a linear combination of monomials of degree n, that is, $P(x) = \sum_{|\alpha|=n} c_\alpha x^\alpha$, where c_α are either real or complex numbers. A polynomial of (total) degree at most n is of the form $P(x) = \sum_{|\alpha| \le n} c_\alpha x^\alpha$. Let $\mathscr{P}_n^d$ denote the space of real homogeneous polynomials of degree n and Π_n^d the space of real polynomials of degree at most n. Counting the cardinalities of $\{\alpha \in \mathbb{N}_0^d : |\alpha| = n\}$ and $\{\alpha \in \mathbb{N}_0^d : |\alpha| \le n\}$ shows that

$$\dim \mathscr{P}_n^d = \binom{n+d-1}{n} \quad \text{and} \quad \dim \Pi_n^d = \binom{n+d}{n}.$$

A harmonic polynomial is a homogeneous polynomial that satisfies the Laplace equation. Let ∂_i be the partial derivative in the i-th variable and Δ the Laplacian operator

$$\Delta = \partial_1^2 + \cdots + \partial_d^2.$$

Definition 1.1.1. For $n = 0,1,2,\ldots$, let $\mathscr{H}_n^d$ be the linear space of real harmonic polynomials, homogeneous of degree n, on $\mathbb{R}^d$, that is,

$$\mathscr{H}_n^d = \left\{ P \in \mathscr{P}_n^d : \Delta P = 0 \right\}.$$

Spherical harmonics are the restrictions of elements in $\mathscr{H}_n^d$ on the unit sphere $\mathbb{S}^{d-1}$. If $Y \in \mathscr{H}_n^d$, then $Y(x) = \|x\|^n Y(x')$, where $x = \|x\| x'$ and $x' \in \mathbb{S}^{d-1}$. We will also call $\mathscr{H}_n^d$ the space of spherical harmonics.

Spherical harmonics of different degrees are orthogonal with respect to the inner product

$$\langle f, g \rangle_{\mathbb{S}^{d-1}} := \frac{1}{\omega_d} \int_{\mathbb{S}^{d-1}} f(x) g(x) d\sigma(x),$$

where $d\sigma$ is the surface area measure on $\mathbb{S}^{d-1}$, and ω_d denotes the surface area of $\mathbb{S}^{d-1}$,

$$\omega_d := \int_{\mathbb{S}^{d-1}} d\sigma = \frac{2\pi^{d/2}}{\Gamma(d/2)}.$$

Theorem 1.1.2. *If $Y_n \in \mathscr{H}_n^d$, $Y_m \in \mathscr{H}_m^d$, and $n \neq m$, then $\langle Y_n, Y_m \rangle_{\mathbb{S}^{d-1}} = 0$. For $n = 0,1,2,\ldots$, $\mathscr{P}_n^d$ admits the decomposition*

$$\mathscr{P}_n^d = \bigoplus_{0 \le j \le n/2} \|x\|^{2j} \mathscr{H}_{n-2j}^d.$$

In other words, for each $P \in \mathscr{P}_n^d$, there is a unique decomposition

$$P(x) = \sum_{0 \le j \le n/2} \|x\|^{2j} P_{n-2j}(x) \quad \text{with} \quad P_{n-2j} \in \mathscr{H}_{n-2j}^d.$$

From the orthogonal decomposition, one immediately deduces the following:

Corollary 1.1.3. *For $n = 0,1,2,\ldots$*

$$\dim \mathscr{H}_n^d = \dim \mathscr{P}_n^d - \dim \mathscr{P}_{n-2}^d = \binom{n+d-1}{n} - \binom{n+d-3}{n-2},$$

with the convention $\dim \mathscr{P}_{n-2}^d = 0$ *for $n = 0,1$.*

In the spherical-polar coordinates $x = r\xi$, $r > 0$, $\xi \in \mathbb{S}^{d-1}$, the Laplace operator is written

$$\Delta = \frac{\partial^2}{\partial r^2} + \frac{d-1}{r} \frac{\partial}{\partial r} + \frac{1}{r^2} \Delta_0,$$

where Δ_0, called the Laplace–Beltrami operator, can be given explicitly by

$$\Delta_0 = \sum_{i=1}^{d-1} \frac{\partial^2}{\partial \xi_i^2} - \sum_{i=1}^{d-1}\sum_{j=1}^{d-1} \xi_i \xi_j \frac{\partial^2}{\partial \xi_i \partial \xi_j} - (d-1) \sum_{i=1}^{d-1} \xi_i \frac{\partial}{\partial \xi_i}.$$

Using this expression of Δ, $\Delta Y = 0$ for Y being a homogeneous polynomial leads to the following result.

Theorem 1.1.4. *The spherical harmonics are eigenfunctions of* Δ_0*:*

$$\Delta_0 Y(\xi) = -n(n+d-2)Y(\xi), \quad \forall Y \in \mathcal{H}_n^d, \quad \xi \in \mathbb{S}^{d-1}.$$

In spherical coordinates, an orthogonal basis of $\mathcal{H}_n^d$ can be given explicitly. Let $\{Y_\alpha\}$ be an orthonormal basis of $\mathcal{H}_n^d$, that is, $\langle Y_\alpha, Y_\beta \rangle_{\mathbb{S}^{d-1}} = \delta_{\alpha,\beta}$. A function f in $L^2(\mathbb{S}^{d-1})$ can be expanded in a Fourier series

$$f(x) = \sum c_\alpha Y_\alpha(x), \quad \text{where} \quad c_\alpha = \frac{1}{\omega_d} \int_{\mathbb{S}^{d-1}} f(y) Y_\alpha(y) d\sigma(y).$$

It is often more convenient to consider the orthogonal expansions in terms of the spaces $\mathcal{H}_n^d$. Collecting terms of spherical harmonics of the same degree, the Fourier series takes the form

$$f(x) = \sum_{n=0}^{\infty} \operatorname{proj}_n f(x),$$

where $\operatorname{proj}_n f$ is the orthogonal projection of f onto the space $\mathcal{H}_n^d$ and satisfies

$$\operatorname{proj}_n f(x) = \frac{1}{\omega_d} \int_{\mathbb{S}^{d-1}} f(y) Z_n(x,y) d\sigma(y), \quad x \in \mathbb{S}^{d-1},$$

in which $Z_n(\cdot,\cdot)$, called the reproducing kernel of $\mathcal{H}_n^d$, is given by

$$Z_n(x,y) = \sum_{k=1}^{\dim \mathcal{H}_n^d} Y_k(x) Y_k(y), \quad x,y \in \mathbb{S}^{d-1}.$$

Since the space of spherical polynomials is dense in $C(\mathbb{S}^{d-1})$ by the Weierstrass theorem and, as a consequence, dense in $L^2(\mathbb{S}^{d-1})$, the following theorem is a standard Hilbert space result for $L^2(\mathbb{S}^{d-1})$:

Theorem 1.1.5. *The set of spherical harmonics is dense in* $L^2(\mathbb{S}^{d-1})$ *and*

$$L^2(\mathbb{S}^{d-1}) = \sum_{n=0}^{\infty} \mathcal{H}_n^d, \quad f = \sum_{n=0}^{\infty} \operatorname{proj}_n f$$

in the sense that $\lim_{n\to\infty} \|f - S_n f\|_2 = 0$ *for any* $f \in L^2(\mathbb{S}^{d-1})$*, where* $S_n f := \sum_{j=0}^{n} \operatorname{proj}_j f$*. In particular, for* $f \in L^2(\mathbb{S}^{d-1})$*, the Parseval identity holds:*

$$\|f\|_2^2 = \sum_{n=0}^{\infty} \|\operatorname{proj}_n f\|_2^2.$$

Much of the analysis on the sphere beyond the L^2 setting depends on the knowledge of the kernel Z_n. It is known that this kernel is uniquely determined by its reproducing property

$$\frac{1}{\omega_d}\int_{\mathbb{S}^{d-1}} Z_n(x,y)p(y)d\sigma(y) = p(x), \quad \forall p \in \mathscr{H}_n^d, \quad x \in \mathbb{S}^{d-1}$$

and the requirement that $Z_n(x,\cdot)$ is an element of $\mathscr{H}_n^d$ for each fixed x. In particular, Z_n is independent of the particular choice of bases of $\mathscr{H}_n^d$. The space $\mathscr{H}_n^d$ is invariant under the action of the orthogonal group $O(d)$ and the surface measure is also invariant, so that the kernel $Z_n(\cdot,\cdot)$ satisfies $Z_n(x,y) = Z_n(xg,yg)$ for all $g \in O(d)$. This implies that $Z_n(x,y)$ depends only on the distance between x and y, where the distance is the geodesic distance $d(x,y) = \arccos\langle x,y\rangle$. Hence, $P_n(x,y) = F_n(\langle x,y\rangle)$, which is often called a zonal harmonic as it is harmonic and depends only on $\langle x,y\rangle$. It turns out that the function F_n has a concise formula in terms of the Gegenbauer polynomial C_n^λ of degree n, defined by

$$C_n^\lambda(x) := \frac{(\lambda)_n 2^n}{n!} x^n {}_2F_1\left(\begin{matrix} -\frac{n}{2}, \frac{1-n}{2} \\ 1-n-\lambda \end{matrix}; \frac{1}{x^2}\right), \tag{1.1.1}$$

for $\lambda > 0$ and $n \in \mathbb{N}_0$, where ${}_2F_1$ is the hypergeometric function.

Theorem 1.1.6. *For $n \in \mathbb{N}_0$ and $x,y \in \mathbb{S}^{d-1}$, $d \geq 3$,*

$$Z_n(x,y) = \frac{n+\lambda}{\lambda} C_n^\lambda(\langle x,y\rangle), \quad \lambda = \frac{d-2}{2}. \tag{1.1.2}$$

The Gegenbauer polynomials are also called ultra-spherical polynomials. They are orthogonal with respect to the weight function

$$w_\lambda(t) := (1-t^2)^{\lambda-\frac{1}{2}}, \quad t \in [-1,1].$$

Let c_λ be the normalization constant of w_λ, $c_\lambda = 1/\int_{-1}^1 w_\lambda(t)dt$. Then

$$c_\lambda \int_{-1}^1 C_n^\lambda(x) C_m^\lambda(x) w_\lambda(x) dx = \frac{\lambda}{(n+\lambda)} \frac{(2\lambda)_n}{n!} \delta_{n,m}.$$

These classical polynomials have been extensively studied. For their essential properties, see [53]. In particular, there is a generating function

$$\frac{1}{(1-2rt+r^2)^{\lambda+1}} = \sum_{n=0}^\infty \frac{n+\lambda}{\lambda} C_n^\lambda(t) r^n, \quad 0 \leq r < 1, \quad \lambda > 0.$$

The concise formula for $Z_n(x,y)$ is one of the most useful ingredients for analysis on the sphere. For example, it leads to the following definition of a convolution on the sphere: for $f \in L^1(\mathbb{S}^{d-1})$ and $g \in L^1(w_\lambda,[-1,1])$ with $\lambda = \frac{d-2}{2}$,

$$(f*g)(x) := \frac{1}{\omega_d}\int_{\mathbb{S}^{d-1}} f(y) g(\langle x,y\rangle) d\sigma(y), \ x \in \mathbb{S}^{d-1}. \tag{1.1.3}$$

The generating function of the Gegenbauer polynomials leads to the following definition: for $f \in L^1(\mathbb{S}^{d-1})$, the Poisson integral of f is

$$P_r f(\xi) := (f * P_r)(\xi), \quad \xi \in \mathbb{S}^{d-1},$$

where the kernel $P_r(t)$ is given by

$$P_r(t) := \frac{1-r^2}{(1-2rt+r^2)^{d/2}}, \quad t \in [-1,1],$$

for $0 \le r < 1$.

The Poisson kernel and Poisson integral satisfy the following properties:

(1) for $x, y \in \mathbb{S}^{d-1}$, $P_r(\langle x,y\rangle) = \sum_{n=0}^{\infty} Z_n(x,y) r^n$;

(2) $P_r f = \sum_{n=0}^{\infty} r^n \operatorname{proj}_n f$;

(3) $P_r(\langle x,y\rangle) \ge 0$ and $\omega_d^{-1} \int_{\mathbb{S}^{d-1}} P_r(\langle x,y\rangle) d\sigma(y) = 1$.

Using these properties, it is easy to prove the following well-known theorem.

Theorem 1.1.7. *Let f be a continuous function on $\mathbb{S}^{d-1}$. For $0 \le r < 1$, $u(r\xi) := P_r f(\xi)$ is a harmonic function in $x = r\xi$, and $\lim_{r \to 1^-} u(r\xi) = f(\xi)$, $\forall \xi \in \mathbb{S}^{d-1}$.*

Spherical harmonics appear in many disciplines and in many different branches of mathematics. We outlined the essential structure for analysis on the sphere. For proofs and further results we refer to [16, 40, 52] and the discussion at the end of Chapter 1 in [16].

1.2 Fourier transform

For $f \in L^1(\mathbb{R}^d)$, the Fourier transform of f is (well) defined by

$$\widehat{f}(x) = \frac{1}{(2\pi)^{d/2}} \int_{\mathbb{R}^d} f(y) e^{-i\langle x,y\rangle} dy, \quad x \in \mathbb{R}^d.$$

For $f \in L^1(\mathbb{R}^d)$, $\widehat{f} \in C_0(\mathbb{R}^d)$. The basic properties of the Fourier transform are summarized in the following theorem:

Theorem 1.2.1. (i) *If $f \in L^1(\mathbb{R}^d)$ and $\widehat{f} \in L^1(\mathbb{R}^d)$, then the inversion formula,*

$$f(y) = \frac{1}{(2\pi)^{d/2}} \int_{\mathbb{R}^d} \widehat{f}(x) e^{-i\langle x,y\rangle} dx,$$

holds for almost every $y \in \mathbb{R}^d$.

(ii) *The Fourier transform extends uniquely to an isometric isomorphism on $L^2(\mathbb{R}^d)$: $\|f\|_2 = \|\widehat{f}\|_2$ for all $f \in L^2(\mathbb{R}^d)$.*

(iii) *If* $f, g \in L^2(\mathbb{R}^d)$, *then*

$$\int_{\mathbb{R}^d} \widehat{f}(x) g(x)\, dx = \int_{\mathbb{R}^d} f(x)\widehat{g}(x) dx.$$

(iv) *If* $f(x) = f_0(\|x\|)$ *is radial, then* $\widehat{f}(x) = H_{\frac{d-2}{2}} f_0(\|x\|)$ *is again a radial function, where* H_α *denotes the Hankel transform defined by*

$$H_\alpha g(s) = \frac{1}{\Gamma(\alpha+1)} \int_0^\infty g(r) \frac{J_\alpha(rs)}{(rs)^\alpha} r^{2\alpha+1}\, dr,$$

in which J_α *denotes the Bessel function of the first kind.*

The usual proof of (i) uses convolution defined by

$$f * g(x) = \frac{1}{(2\pi)^{d/2}} \int_{\mathbb{R}^d} f(y) g(x-y) dy,$$

for $f, g \in L^1(\mathbb{R}^d)$. It is easy to see that if $f, g \in L^1(\mathbb{R}^d)$, then

$$\widehat{f * g}(x) = \widehat{f}(x)\widehat{g}(x).$$

Let Φ be a nice function, say $\Phi(x) = e^{-\|x\|}$ or $e^{-\|x\|^2/2}$, and let $\phi := \widehat{\Phi}$; normalize Φ so that $\int_{\mathbb{R}^d} \phi(x) = 1$. For $\varepsilon > 0$, define $\phi_\varepsilon(x) := \varepsilon^{-d}\phi(x/\varepsilon)$. It is easy to see that

$$(f * \phi_\varepsilon)(x) = \frac{1}{(2\pi)^{d/2}} \int_{\mathbb{R}^d} \Phi(\varepsilon y) \widehat{f}(y) e^{i\langle x,y\rangle} dy.$$

Thus, the proof of (i) comes down to showing that $f * \phi_\varepsilon(x) \to f(x)$ as $\varepsilon \to 0$.

The eigenfunctions of the Fourier transform can be given in terms of spherical harmonics. Let $Y \in \mathcal{H}_n^d$. Define

$$\phi_m(Y; x) = L_m^{n+\frac{d-2}{2}}(\|x\|^2) Y(x) e^{-\|x\|^2/2}, \qquad x \in \mathbb{R}^d,$$

where L_n^α denotes the Laguerre polynomial of degree n with index α, normalized so that

$$\frac{1}{\Gamma(\alpha+1)} \int_0^\infty L_n^\alpha(x) L_m^\alpha(x) x^\alpha e^{-x} dx = \binom{n+\alpha}{\alpha} \delta_{m,n}.$$

If $\{Y_{k,n} : 1 \le k \le \dim \mathcal{H}_n^d\}$ denotes an orthonormal basis of $\mathcal{H}_n^d$, then it is easy to verify, using spherical polar coordinates and the orthogonality of L_n^α, that $\{\phi_m(Y_{k,n}; x) : m, n \ge 0, 1 \le k \le \dim \mathcal{H}_n^d\}$ is an orthogonal basis of $L^2(\mathbb{R}^d)$.

Theorem 1.2.2. *For* $m, n = 0, 1, 2, \ldots$, $Y \in \mathcal{H}_n^d$ *and* $x \in \mathbb{R}^d$,

$$\widehat{\phi_m(Y)}(x) = (-i)^{n+2m} \phi_m(Y; x).$$

There are many books on Fourier transforms. For the basics that we need here and the proofs, we refer to [31, 46, 52].

Chapter 2

Dunkl Operators Associated with Reflection Groups

In this chapter we introduce the essential ingredient in the Dunkl theory of harmonic analysis. Since our purpose is to study harmonic analysis in weighted spaces, we start with the definition of a family of weight functions invariant under a reflection group in Section 2.1. Dunkl operators are a family of commuting first-order differential and difference operators associated with a reflection group, and are introduced in Section 2.2. The intertwining operator between the Dunkl operators and ordinary derivatives is discussed in Section 2.3.

For readers who are primarily interested in analysis, the prerequisites on reflection groups are reduced to a minimum. In fact, all essential ideas are presented in the case of $G = \mathbb{Z}_2^d$, which requires no prior knowledge of reflection groups.

2.1 Weight functions invariant under a reflection group

The simplest family of weight functions in d variables that we consider is defined by

$$h_\kappa(x) := \prod_{i=1}^{d} |x_i|^{\kappa_i}, \quad x \in \mathbb{R}^d, \tag{2.1.1}$$

for $\kappa_i \ge 0$, $1 \le i \le d$, and $x = (x_1, \ldots, x_d)$. Obviously, they are invariant under sign changes, that is, invariant under the group $\mathbb{Z}_2^d$. This is a special case of weight functions invariant under reflection groups. To define the general weight functions, we first need to recall basic facts on reflection groups. Readers who are not interested in reflection groups can skip to the end of the section and keep in mind the functions h_κ in (2.1.1) and $\mathbb{Z}_2^d$ in the rest of these lecture notes.

For $x \in \mathbb{R}^d$, let $\langle x, y \rangle$ denote the usual Euclidean inner product and $\|x\| := \sqrt{\langle x, x \rangle}$ the Euclidean norm of x. For a nonzero vector $v \in \mathbb{R}^d$, let σ_v denote the reflection with respect to the hyperplane $v^\perp$ perpendicular to v,

$$x\sigma_v := x - 2(\langle x,v\rangle/\|v\|^2)v, \quad x \in \mathbb{R}^d.$$

A root system is a finite set R of nonzero vectors in $\mathbb{R}^d$ such that $u,v \in R$ implies $u\sigma_v \in R$. If, in addition, $u,v \in R$ and $u = cv$ for some scalar c implies that $c = \pm 1$, then R is called reduced. The set $\{u^\perp : u \in R\}$ is a finite set of hyperplanes, hence, there exists $u_0 \in \mathbb{R}^d$ such that $\langle u,u_0\rangle \neq 0$ for all $u \in R$. With respect to u_0 define the set of positive roots $R_+ := \{v \in R : \langle v,u_0\rangle > 0\}$. If $u \in R$, then $-u = u\sigma_u \in R$, so that $R = R_+ \cup (-R_+)$.

The finite reflection group G generated by the root system R is the subgroup of $O(d)$ generated by $\{\sigma_u : u \in R\}$. If R is reduced, then the set of reflections contained in G is exactly $\{\sigma_u : u \in R_+\}$. For a given root system R, a multiplicity function $v \mapsto \kappa_v : R \to \mathbb{R}_{\geq 0}$ is a nonnegative function defined on R such that $\kappa_v = \kappa_u$ whenever σ_u is conjugate to σ_v, that is, there exists $g \in G$ such that $ug = v$.

Given a reduced root system R on $\mathbb{R}^d$ and a multiplicity function κ_v on R, we define a weight function h_κ by

$$h_\kappa(x) := \prod_{v\in R_+} |\langle x,v\rangle|^{\kappa_v}, \quad x \in \mathbb{R}^d. \tag{2.1.2}$$

Then h_κ is invariant under the reflection group G generated by R. It is a homogeneous function of degree

$$\gamma_\kappa := \sum_{v\in R_+} \kappa_v. \tag{2.1.3}$$

For h_κ in (2.1.1) associated with $\mathbb{Z}_2^d$, $\gamma_\kappa = |\kappa| = \kappa_1 + \cdots + \kappa_d$.

Let us give two examples beyond $\mathbb{Z}_2^d$. Let $e_1,\ldots,e_d$ be the standard Euclidean basis, that is, the i-th component of e_i is 1 and all other components are 0.

Symmetric group. The root system is $R = \{e_i - e_j : 1 \leq i \neq j \leq d\}$. Choosing $u_0 = (d, d-1,\ldots,1)$, one has $R_+ = \{e_i - e_j : 1 \leq i < j \leq d\}$. There is only one conjugacy class in this group, so that the weight function is

$$h_\kappa(x) = \prod_{1\leq i<j\leq d} |x_i - x_j|^\kappa, \quad \kappa \geq 0, \quad x \in \mathbb{R}^d. \tag{2.1.4}$$

This reflection group is of the type A_{d-1} and it is the same as the symmetric, or permutation, group of d objects. Evidently, h_κ is symmetric under permutations of $x_1,\ldots,x_d$.

Octahedral group. The positive root system is $R_+ = \{e_i - e_j, e_i + e_j : 1 \leq i \neq j \leq d\} \cup \{e_i : 1 \leq i \leq d\}$. There are two conjugacy classes in this group, so that the weight function is

$$h_\kappa(x) = \prod_{i=1}^d |x_i|^{\kappa_0} \prod_{1\leq i<j\leq d} |x_i^2 - x_j^2|^{\kappa_1}, \quad \kappa_0,\kappa_1 \geq 0, \quad x \in \mathbb{R}^d. \tag{2.1.5}$$

This reflection group is of the type B_d and it is the symmetric group of the octahedron $\{\pm e_1,\ldots,\pm e_d\}$ of $\mathbb{R}^d$, or the cube in $\mathbb{R}^d$. Obviously, h_κ is symmetric under permutations of $x_1,\ldots,x_d$ and sign changes.

The analysis in these lecture notes is in the setting of weighted L^p spaces with these reflection invariant weight functions on the unit sphere and on $\mathbb{R}^d$.

Let $d\sigma$ denote the surface measure on the unit sphere $\mathbb{S}^{d-1}$. Let ω_d denote the surface area and ω_d^κ denote the normalization constant of h_κ:

$$\omega_d := \int_{\mathbb{S}^{d-1}} d\sigma = \frac{2\pi^{d/2}}{\Gamma(d/2)} \quad \text{and} \quad \omega_d^\kappa := \int_{\mathbb{S}^{d-1}} h_\kappa^2(y) d\sigma.$$

The closed form of ω_d^κ is known for every reflection group (cf. [26]). For $1 \le p \le \infty$ we denote by $L^p(h_\kappa^2)$ the space of functions defined on $\mathbb{S}^{d-1}$ with finite norm

$$\|f\|_{\kappa,p} := \left(\frac{1}{\omega_d^\kappa} \int_{\mathbb{S}^{d-1}} |f(y)|^p h_\kappa^2(y) d\sigma(y) \right)^{1/p}, \qquad 1 \le p < \infty,$$

and for $p = \infty$ we assume that L^∞ is replaced by $C(\mathbb{S}^{d-1})$, the space of continuous functions on $\mathbb{S}^{d-1}$ with the usual uniform norm $\|f\|_\infty$.

Let $\mathscr{P}_n^d$ denote the space of homogeneous polynomials of degree n in d variables. Consider the measure on $\mathbb{R}^d$ defined by

$$d\mu := h_\kappa^2(x) e^{-\|x\|^2/2} dx, \qquad x \in \mathbb{R}^d,$$

and let c_h denote the normalization constant

$$c_h := \left(\frac{1}{(2\pi)^{d/2}} \int_{\mathbb{R}^d} h_\kappa^2(x) e^{-\|x\|^2/2} dx \right)^{-1}. \tag{2.1.6}$$

The inner product with respect to $d\mu$ is closely related to the one with respect to dx when we restrict to the space of homogeneous polynomials. For $n \in \mathbb{N}_0$ and $a \in \mathbb{R}$, let $(a)_n := a(a+1)\cdots(a+n-1)$ denote the shifted factorial of a.

Proposition 2.1.1. *For $p, q \in \mathscr{P}_n^d$,*

$$c_h \int_{\mathbb{R}^d} p(x) q(x) h_\kappa^2(x) e^{-\|x\|^2/2} dx = (2\pi)^{d/2} 2^n (\lambda_\kappa + 1)_n \frac{1}{\omega_d^\kappa} \int_{\mathbb{S}^{d-1}} p(\xi) q(\xi) h_\kappa^2(\xi) d\sigma(\xi),$$

where $\lambda_\kappa = \frac{d-2}{2} + \gamma_\kappa$.

Proof. If g is a homogeneous polynomial of degree $2n$, then using spherical polar coordinates $x = r\xi$, $r > 0$ and $\xi \in \mathbb{S}^{d-1}$, we have

$$\begin{aligned} \int_{\mathbb{R}^d} g(x) h_\kappa^2(x) e^{-\|x\|^2/2} dx &= \int_0^\infty r^{2\gamma_\kappa + 2n + d - 1} e^{-r^2/2} dr \int_{\mathbb{S}^{d-1}} g(\xi) h_\kappa^2(\xi) d\sigma(\xi) \qquad (2.1.7) \\ &= 2^{n+\lambda_\kappa} \Gamma(\lambda_\kappa + n + 1) \int_{\mathbb{S}^{d-1}} g(\xi) h_\kappa^2(\xi) d\sigma(\xi), \end{aligned}$$

from which the result follows. □

It is worth noting that, setting $n = 0$ in (2.1.7),

$$c_h^{-1} = \frac{1}{(2\pi)^{d/2}} 2^{\lambda_\kappa} \Gamma(\lambda_\kappa + 1) \omega_d^\kappa = 2^{\lambda_\kappa} \frac{\Gamma(\lambda_\kappa + 1)}{\Gamma(d/2)} \frac{\omega_d^\kappa}{\omega_d}. \tag{2.1.8}$$

2.2 Dunkl operators

The main ingredient of the theory of h-harmonics is a family of first-order differential-difference operators, $\mathscr{D}_i$, called the Dunkl operators.

Definition 2.2.1. Let R_+ be a positive root system and κ_v be a multiplicity function from R_+ to $\mathbb{R}_{\geq 0}$. For $1 \leq i \leq d$, define

$$\mathscr{D}_i f(x) := \partial_i f(x) + \sum_{v \in R_+} \kappa_v \frac{f(x) - f(x\sigma_v)}{\langle x, v\rangle} \langle v, e_i\rangle, \qquad 1 \leq i \leq d. \tag{2.2.1}$$

It is easy to verify that $\mathscr{D}_i \mathscr{P}_n^d \subset \mathscr{P}_{n-1}^d$, so that the $\mathscr{D}_i$ are indeed first-order differential-difference operators. In the case of $\mathbb{Z}_2^d$, the Dunkl operators take on the form

$$\mathscr{D}_i f(x) = \partial_i f(x) + \kappa_i \frac{f(x) - f(x\sigma_i)}{x_i} \tag{2.2.2}$$

where $x\sigma_i = (x_1, \ldots, x_{i-1}, -x_i, x_{i+1}, \ldots, x_d)$.

The most important property of these operators is that they commute.

Theorem 2.2.2. *The Dunkl operators commute:*

$$\mathscr{D}_i \mathscr{D}_j = \mathscr{D}_j \mathscr{D}_i, \qquad 1 \leq i, j \leq d.$$

Proof. The proof of the general case is rather involved. We give the proof only for the case of $\mathbb{Z}_2^d$, for which a straightforward computation shows that, for $i \neq j$,

$$\begin{aligned}\mathscr{D}_i \mathscr{D}_j f(x) = \partial_i \partial_j f(x) &+ \frac{\kappa_i}{x_i}(\partial_j f(x) - \partial_j f(x\sigma_i)) + \frac{\kappa_j}{x_j}(\partial_i f(x) - \partial_i f(x\sigma_j)) \\ &+ \frac{\kappa_i \kappa_j}{x_i x_j}(f(x) - f(x\sigma_j) - f(x\sigma_i) + f(x\sigma_j\sigma_i)),\end{aligned}$$

from which $\mathscr{D}_i \mathscr{D}_j = \mathscr{D}_j \mathscr{D}_i$ follows immediately. □

The Dunkl operators are akin to the partial derivatives and they can be used to define an analog of the Laplace operator, denoted by Δ_h:

$$\Delta_h := \mathscr{D}_1^2 + \cdots + \mathscr{D}_d^2. \tag{2.2.3}$$

This is a second-order differential-difference operator and it reduces to the usual Laplacian Δ when all $\kappa_i = 0$. It has the following explicit formula related to the weight function h_κ in (2.1.2):

Proposition 2.2.3. *The Dunkl Laplacian Δ_h can be written as $\Delta_h = D_h + E_h$, with*

$$D_h f := \frac{\Delta(f h_k) - f \Delta h_k}{h_\kappa} \quad \text{and} \quad E_h f := -2 \sum_{v \in R_+} \kappa_v \frac{f(x) - f(x\sigma_v)}{\langle v, x\rangle^2} \|v\|^2,$$

and both D_h and E_h commute with the action of the reflection group.

Proof. Again, we give the proof only for the case of $\mathbb{Z}_2^d$. Let

$$E_j f(x) := \frac{f(x) - f(x\sigma_j)}{x_j}, \qquad 1 \le j \le d,$$

so that $\mathscr{D}_j = \partial_j + \kappa_j E_j$ in the case of $G = \mathbb{Z}_2^d$. A straightforward computation shows that $E_j^2 = 0$ and

$$\mathscr{D}_j^2 = \partial_j^2 f + \kappa_j \partial_j E_j + \kappa_j E_j \partial_j = \partial_j^2 + 2\frac{\kappa_j}{x_j}\partial_j - \frac{\kappa_j}{x_j}E_j.$$

Summing over j we obtain

$$\mathscr{D}_h = \Delta f + 2\sum_{j=1}^{d} \frac{\kappa_j}{x_j}\partial_j f - \sum_{j=1}^{d} \frac{\kappa_j}{x_j}E_j.$$

In this case the sum over R_+ means the sum over $1 \le j \le d$, so that the sum over E_j gives E_h and the differential part is D_h, which can be written in the stated expression in terms of h_κ by a simple verification. □

Later we will need to perform an integration by parts for the Dunkl operator, at least over the space of polynomials. For this to make sense, we consider the integral with respect to the measure

$$d\mu := h_\kappa^2(x) e^{-\|x\|^2/2} dx, \qquad x \in \mathbb{R}^d.$$

Theorem 2.2.4. *The adjoint $\mathscr{D}_i^*$ acting on $L^2(\mathbb{R}^d; d\mu)$ is given by*

$$\mathscr{D}_i^* p(x) = x_i p(x) - \mathscr{D}_i p(x), \qquad p \in \Pi^d.$$

Proof. Assume $\kappa_v \ge 1$. Analytic continuation can be used to extend the range of validity to $\kappa_v \ge 0$. Let p and q be two polynomials. Integrating by parts shows that

$$\begin{aligned}\int_{\mathbb{R}^d} \big(\partial_i p(x)\big) q(x) h_\kappa^2(x) e^{-\|x\|^2/2} dx = &- \int_{\mathbb{R}^d} p(x) \big(\partial_i q(x)\big) h_\kappa^2(x) e^{-\|x\|^2/2} dx \\ &+ \int_{\mathbb{R}^d} p(x) q(x) \big[-2h_\kappa(x)\partial_i h_\kappa(x) + h_\kappa^2(x) x_i\big] e^{-\|x\|^2/2} dx.\end{aligned}$$

For a fixed root v,

$$\begin{aligned}\int_{\mathbb{R}^d} \frac{p(x) - p(x\sigma_v)}{\langle x, v\rangle} q(x) d\mu &= \int_{\mathbb{R}^d} \frac{p(x)q(x)}{\langle x, v\rangle} d\mu - \int_{\mathbb{R}^d} \frac{p(x\sigma_v)q(x)}{\langle x, v\rangle} d\mu \\ &= \int_{\mathbb{R}^d} \frac{p(x)q(x)}{\langle x, v\rangle} d\mu + \int_{\mathbb{R}^d} \frac{p(x)q(x\sigma_v)}{\langle x, v\rangle} d\mu,\end{aligned}$$

where in the second integral we have replaced x by $x\sigma_v$ which changes $\langle x, v\rangle$ to $\langle x\sigma_v, v\rangle = -\langle x, v\rangle$ and leaves h_κ^2 invariant. Note also that

$$h_\kappa(x)\partial_i h_\kappa(x) = \sum_{v \in R_+} \kappa_v \frac{v_i}{\langle x, v\rangle} h_\kappa^2(x).$$

Combining these ingredients, we obtain

$$\int_{\mathbb{R}^d} \mathscr{D}_i p(x) q(x) d\mu = \int_{\mathbb{R}^d} \Big[p(x)\big(x_i q(x) - \partial_i q(x)\big) + \sum_{v \in R_+} \big(\kappa_v v_i\, p(x)\big(-2q(x) + q(x) + q(x\sigma_v)\big) / \langle x, v\rangle\big) \Big] d\mu,$$

where the term inside the square brackets is exactly $p(x)\big(x_i q(x) - \mathscr{D}_i q(x)\big)$. □

2.3 Intertwining operator

There is a linear operator that intertwines between the Dunkl operators and the partial derivatives, which plays an important role in harmonic analysis.

Definition 2.3.1. Let $\mathscr{D}_i$ be the Dunkl operators associated with a given positive root system and a multiplicity function κ. A linear operator V_κ on the space Π^d of algebraic polynomials on $\mathbb{R}^d$ is called an intertwining operator if it satisfies

$$\mathscr{D}_i V_\kappa = V_\kappa \partial_i, \quad 1 \le i \le d, \quad V_\kappa 1 = 1, \quad V_\kappa \mathscr{P}_n \subset \mathscr{P}_n, \ n \in \mathbb{N}_0. \tag{2.3.1}$$

Strictly speaking, (2.3.1) is not the definition of V_κ, but rather the property that we most often use. Indeed, the existence of such a V_κ is by no means automatic. The operator V_κ was introduced in [26], where it was defined inductively on homogeneous polynomials. The definition is extended from homogeneous polynomials to the space $A(\mathbb{B}^d)$ defined below.

For $f \in \Pi^d$, let $\|\cdot\|_A := \sum_{n=0}^\infty \|f_n\|_S$, where $f = \sum_{n=0}^\infty f_n$ with $f_n \in \mathscr{P}_n^d$ and $\|f\|_S = \sup_{x \in \mathbb{S}^{d-1}} |f(x)|$. Let $A(\mathbb{B}^d)$ be the closure of Π^d in A-norm. Then $A(\mathbb{B}^d)$ is a commutative Banach algebra under the pointwise operations and it is contained in $C(\mathbb{B}^d) \cap C^\infty(\{x : \|x\| < 1\})$, where $\mathbb{B}^d = \{x : \|x\| \le 1\}$ is the unit ball of $\mathbb{R}^d$. Then the following proposition holds (see [26]):

Proposition 2.3.2. *For $f \in \Pi^d$ and $x \in \mathbb{B}^d$,*

$$|V_\kappa f(x)| \le \|f\|_A \quad \text{and} \quad \|V_\kappa f\|_A \le \|f\|_A.$$

The existence of the operator V_κ for a generic reflection group satisfying (2.3.1) requires substantial knowledge of reflection groups and considerable efforts [26]. For our purpose, however, it is not necessary to know the proof.

In the case of $\mathbb{Z}_2^d$, the intertwining operator V_κ has an explicit expression as an integral operator.

Theorem 2.3.3. *Let $\kappa_i \ge 0$. The intertwining operator for $\mathbb{Z}_2^d$ is given by*

$$V_\kappa f(x) = c_\kappa \int_{[-1,1]^d} f(x_1 t_1, \ldots, x_d t_d) \prod_{i=1}^d (1 + t_i)(1 - t_i^2)^{\kappa_i - 1} dt_i, \tag{2.3.2}$$

where $c_\kappa = c_{\kappa_1} \cdots c_{\kappa_d}$ *with* $c_\mu = \Gamma(\mu + 1/2)/(\sqrt{\pi}\Gamma(\mu))$, *and if any one of* $\kappa_i = 0$, *then the formula holds under the limit*

$$\lim_{\mu \to 0} c_\mu \int_{-1}^{1} f(t)(1-t^2)^{\mu-1} dt = \frac{f(1)+f(-1)}{2}. \tag{2.3.3}$$

Proof. The integrals are normalized so that $V_\kappa 1 = 1$. Recall that $\mathscr{D}_j = \partial_j + \kappa_j E_j$. Taking derivatives we get

$$\partial_j V_\kappa f(x) = c_\kappa \int_{[-1,1]^d} \partial_j f(x_1 t_1, \ldots, x_d t_d) t_j \prod_{i=1}^{d} (1+t_i)(1-t_i^2)^{\kappa_i - 1} dt_i.$$

Taking into account the parity of the integrand, integration by parts shows that

$$\begin{aligned} \kappa_j E_j V_\kappa f(x) &= \frac{\kappa_j}{x_j} c_\kappa \int_{[-1,1]^d} f(x_1 t_1, \ldots, x_d t_d) 2t_j \Big(\prod_{i \neq j} (1+t_i)\Big) \Big(\prod_{i=1}^{d} (1-t_i^2)^{\kappa_i - 1}\Big) dt_i \\ &= c_\kappa \int_{[-1,1]^d} \partial_j f(x_1 t_1, \ldots, x_d t_d)(1-t_j) \prod_{i=1}^{d} (1+t_i)(1-t_i^2)^{\kappa_i - 1} dt_i. \end{aligned}$$

Adding the last two equations gives $\mathscr{D}_j V_\kappa = V_\kappa \partial_j$ for $1 \le j \le d$. □

Apart from partial results for the symmetric group on three variables and the dihedral group D_4, an explicit formula for V_κ is not known. In general, however, we have the following theorem of Rösler, which, in particular, asserts that V_κ is nonnegative.

Theorem 2.3.4. *For each* $x \in \mathbb{R}^d$, *there exists a unique probability measure* μ_x^κ *on the Borel* σ*-algebra of* $\mathbb{R}^d$ *such that for all algebraic polynomials* f *on* $\mathbb{R}^d$,

$$V_\kappa f(x) = \int_{\mathbb{R}^d} f(\xi) d\mu_x^\kappa(\xi). \tag{2.3.4}$$

Furthermore, the measures μ_x^κ *are compactly supported in the convex hull* $C(x) := \mathrm{conv}\{xg : g \in G\}$ *of the orbit of* x *under* G, *and satisfy*

$$\mu_{rx}^\kappa(E) = \mu_x^\kappa(r^{-1}E), \ \text{and}\ \mu_{xg}^\kappa(E) = \mu_x^\kappa(Eg^{-1}) \tag{2.3.5}$$

for all $r > 0$, $g \in G$ *and each Borel subset* E *of* $\mathbb{R}^d$.

Remark 2.3.5. This theorem was proved in [43, Th. 1.2 and Cor. 5.3]. Note that the measure μ_x^κ depends on x. The most useful part of the result is that V_κ is a nonnegative operator. By means of (2.3.4), V_κ can be extended to an operator in the space $C(\mathbb{R}^d)$ of continuous functions on $\mathbb{R}^d$, which we will denote by V_κ again.

Definition 2.3.6. For $x, y \in \mathbb{R}^d$, define

$$E(x,y) := V_\kappa^{(x)}\left(e^{\langle x,y \rangle}\right)$$

and, for $n = 0, 1, 2, \ldots$, define

$$E_n(x,y) := \frac{1}{n!} V_\kappa^{(x)}(\langle x,y\rangle^n),$$

where the superscript x means that V_κ acts on the x variable.

Definition 2.3.7. For $p, q \in \mathscr{P}_n^d$, define $\langle p,q\rangle_{\mathscr{D}} := p(\mathscr{D})q(x)$.

Proposition 2.3.8. *The kernel E_n satisfies the following properties*

(1) *E_n is symmetric, $E_n(x,y) = E_n(y,x)$;*

(2) *$E_n(xg,yg) = E_n(x,y)$, $g \in G$;*

(3) *E_n is the reproducing kernel of $\langle\cdot,\cdot\rangle_{\mathscr{D}}$, that is,*

$$\langle E_n(x,\cdot), p\rangle_{\mathscr{D}} = p(x), \qquad \forall p \in \mathscr{P}_n^d.$$

Proof. The first two properties follow from the inductive definition of V_κ, see [26]. For the third property, we note that if $p \in \mathscr{P}_n^d$, then $p(x) = (\langle x, \partial^{(y)}\rangle^n/n!)p(y)$. Applying $V_\kappa^{(x)}$ leads to $V_\kappa^{(x)} p(x) = E_n(x, \partial^{(y)})p(y)$. The left-hand side is independent of y so, applying $V^{(y)}$ to both sides, we get $V_\kappa^{(x)} p(x) = E_n(x, \mathscr{D}^{(y)})V_\kappa^{(y)} p(y)$. Thus, the desired identity holds for all $V_\kappa p$ with $p \in \mathscr{P}_n^d$, which completes the proof, since V_κ is one-to-one. □

2.4 Notes and further results

The Dunkl operators were introduced in [25] and the intertwining operators and the inner products in Section 2.3 were studied in [26]. For a complete proof of the existence of the intertwining operator and its basic properties, see [29, Chapter 6]. The positivity of the intertwining operator was proved in [43]. The explicit formula for V_κ in the case of $\mathbb{Z}_2^d$ was given in [69]. For the symmetric group S_3 with $h_\kappa(x) = |(x_1-x_2)(x_2-x_3)(x_3-x_1)|^\kappa$ for $x \in \mathbb{S}^2$ and the dihedral group $I(4)$ with $h_\kappa(x) = |x_1x_2|^{\kappa_0}|x_1^2 - x_2^2|^{\kappa_1}$, some explicit integral formulas for V_κ are given in [28] and [70], respectively. But neither of them is in a strong enough form for carrying out the harmonic analysis that will be developed in latter chapters.

Chapter 3

h-Harmonics and Analysis on the Sphere

Dunkl h-harmonics are defined as homogeneous polynomials satisfying the Dunkl Laplacian equation. They are defined and studied in Section 3.1. Projection operators and orthogonal expansions in spherical h-harmonics are studied in Section 3.2, which includes a concise expression for the reproducing kernel of the spherical h-harmonics. This expression is an analog of the zonal harmonics, which suggests a definition of a convolution operator, defined in Section 3.3 and it helps us to study various summability methods for spherical h-harmonic expansions. Maximal functions are introduced in Section 3.4 and proved to be of strong type (p,p) and weak type $(1,1)$. Finally, the relation between convolution and maximal functions is discussed in Section 3.5.

3.1 Dunkl h-harmonics

We are now in a position to define h-harmonics.

Definition 3.1.1. Let $Y \in \mathscr{P}_n^d$ be a homogeneous polynomial of degree n. If $\Delta_h Y = 0$, then Y is called an h-harmonic polynomial of degree n.

For $n = 0, 1, 2, \ldots$, let $\mathscr{H}_n^d(h_\kappa^2)$ denote the linear space of h-harmonic polynomials of degree n. Elements of $\mathscr{H}_n^d(h_\kappa^2)$ are homogeneous polynomials so that they are uniquely determined by their restrictions to the unit sphere $\mathbb{S}^{d-1}$. The restrictions of h-harmonics to the sphere are spherical h-harmonics, analogues to spherical harmonics. We shall not distinguish between $Y_n^h \in \mathscr{H}_n^d(h_\kappa^2)$ and its restriction to the sphere.

Let h_κ be the weight function defined in (2.1.2). The inner product in $L^2(h_\kappa^2, \mathbb{S}^{d-1})$ is denoted by

$$\langle f, g \rangle_\kappa := \frac{1}{\omega_d^\kappa} \int_{\mathbb{S}^{d-1}} f(x) g(x) h_\kappa^2(x) d\sigma(x). \tag{3.1.1}$$

Theorem 3.1.2. *With respect to $\langle\cdot,\cdot\rangle_\kappa$, spherical h-harmonics of different degree are orthogonal. More precisely, if $f \in \mathscr{H}_n^d(h_\kappa^2)$, $g \in \mathscr{H}_m^d(h_\kappa^2)$ and $n \neq m$, then $\langle f,g\rangle_\kappa = 0$.*

Proof. As in the classical proof for ordinary harmonics, this follows from an analog of Green's formula stated for the differentiation part D_h of Δ_h:

$$\int_{\mathbb{S}^{d-1}} \frac{\partial f}{\partial n} g h_\kappa^2 d\sigma = \int_{\mathbb{B}^d} (gD_h f + \langle \nabla f, \nabla g\rangle) h_\kappa^2 dx,$$

where $\partial f/\partial n$ denotes the normal derivative of f. Consequently, since $\frac{\partial f}{\partial n} = nf$ for f homogeneous of degree n, and $\Delta_h f = 0$, $\Delta_h g = 0$,

$$\begin{aligned}(n-m)\int_{\mathbb{S}^{d-1}} fgh_\kappa^2 d\sigma &= \int_{\mathbb{B}^d} (gD_h f - fD_h g)h_\kappa^2 dx = -\int_{\mathbb{B}^d} (gE_h f - fE_h g)h_\kappa^2 dx \\ &= -\int_0^1 r^{2\gamma_\kappa+n+m+d-5} dr \int_{\mathbb{S}^{d-1}} (gE_h f - fE_h g) h_\kappa^2 d\sigma,\end{aligned}$$

using the spherical polar coordinates $x = r\xi$, $r > 0$ and $\xi \in \mathbb{S}^{d-1}$. The last integral is zero since the difference part E_h of Δ_h is self-adjoint with respect to $\langle\cdot,\cdot\rangle_\kappa$. □

Theorem 3.1.3. *For $n = 0,1,2,\ldots$, $\mathscr{P}_n^d$ admits the decomposition*

$$\mathscr{P}_n^d = \bigoplus_{0\le j\le n/2} \|x\|^{2j} \mathscr{H}_{n-2j}^d(h_\kappa^2). \tag{3.1.2}$$

Furthermore, for $n = 0,1,2,\ldots$,

$$\dim \mathscr{H}_n^d(h_\kappa^2) = \dim \mathscr{P}_n^d - \dim \mathscr{P}_{n-2}^d = \binom{n+d-1}{d-1} - \binom{n+d-3}{d-1}. \tag{3.1.3}$$

Proof. Briefly, the proof follows by induction, using the orthogonality of $\mathscr{H}_n^d(h_\kappa^2)$ and the fact that Δ_h maps $\mathscr{P}_n^d$ onto $\mathscr{P}_{n-2}^d$. □

From (2.3.1) it follows immediately that $\Delta_h V_\kappa = V_\kappa \Delta$ and, consequently, if P is an ordinary harmonic polynomial, then $V_\kappa P$ is an h-harmonic.

In terms of the spherical polar coordinates $x = r\xi$, the Dunkl Laplacian Δ_h admits a decomposition as in the case of ordinary Laplace operator. Let us define

$$\lambda_k := \gamma_k + \frac{d-2}{2} = \sum_{v\in R_+} \kappa_v + \frac{d-2}{2}. \tag{3.1.4}$$

Lemma 3.1.4. *In the spherical-polar coordinates $x = r\xi$, $r > 0$, $\xi \in \mathbb{S}^{d-1}$, the Dunkl Laplace operator can be expressed as*

$$\Delta_h = \frac{d^2}{dr^2} + \frac{2\lambda_\kappa + 1}{r}\frac{d}{dr} + \frac{1}{r^2}\Delta_{h,0}, \tag{3.1.5}$$

where

$$\Delta_{h,0} f = \frac{1}{h_\kappa}\left[\Delta_0(f h_\kappa) - f\Delta_0 h_\kappa\right] - E_h^{(\xi)} f, \tag{3.1.6}$$

Δ_0 denotes the usual Laplace–Beltrami operator, and $E_h^{(\xi)}$ means that E_h is acting on the ξ variable.

Proof. By the decomposition $\Delta_h = D_h + E_h$, we can apply the decomposition of the ordinary Laplacian Δ to the differential part D_h, which gives the part of $\Delta_{h,0}$ expressed in terms of the classical Laplace–Beltrami operator Δ_0. The difference part follows readily from the definition of E_h. □

The operator $\Delta_{h,0}$ is the analogue of the Laplace–Beltrami operator on the sphere, which, in particular, has spherical h-harmonics as eigenfunctions.

Theorem 3.1.5. *The spherical h-harmonics are eigenfunctions of $\Delta_{h,0}$:*

$$\Delta_{h,0} Y_n^h(\xi) = -n(n+2\lambda_\kappa) Y_n^h(\xi), \quad \forall Y_n^h \in \mathscr{H}_n^d(h_\kappa^2), \quad \xi \in \mathbb{S}^{d-1}. \tag{3.1.7}$$

Proof. Since Y_n^h is a homogeneous polynomial of degree n, $Y_n^h(x) = r^n Y_n^h(\xi)$. Applying (3.1.5) to Y_n^h, equation (3.1.7) follows from $\Delta_h Y_n^h = 0$. □

The following theorem gives an orthogonal basis for $\mathscr{H}_n^d(h_\kappa^2)$.

Theorem 3.1.6. *For $\alpha \in \mathbb{N}_0^d$, $n = |\alpha|$, define*

$$p_\alpha(x) := \frac{(-1)^n}{2^n(\lambda_\kappa)_n} \|x\|^{2|\alpha|+2\lambda_\kappa} \mathscr{D}^\alpha \{\|x\|\}^{-2\lambda_k}, \tag{3.1.8}$$

where $\mathscr{D}^\alpha := \mathscr{D}_d^{\alpha_d} \cdots \mathscr{D}_1^{\alpha_1}$. Then

1. *$p_\alpha \in \mathscr{H}_n^d(h_\kappa^2)$ and p_α is a monic spherical h-harmonic of the form*

$$p_\alpha(x) = x^\alpha + \|x\|^2 q_\alpha(x), \qquad q_\alpha \in \mathscr{P}_{n-2}^d; \tag{3.1.9}$$

2. *p_α satisfies the recurrence relation*

$$p_{\alpha+e_i}(x) = x_i p_\alpha(x) - \frac{1}{2n+2\lambda_\kappa}\|x\|^2 \mathscr{D}_i p_\alpha(x);$$

3. *$\{p_\alpha : |\alpha| = n, \alpha_d = 0$ or $1\}$ is a basis of $\mathscr{H}_n^d(h_\kappa^2)$.*

The proof of this theorem is more or less a straightforward computation. Indeed, for $g \in \mathscr{P}_n^d$ and $\rho \in \mathbb{R}$, the explicit expressions for $\mathscr{D}_i$ and Δ_h can be used to show that

$$\mathscr{D}_i(\|x\|^\rho g) = \rho x_i \|x\|^{\rho-2} g + \|x\|^\rho \mathscr{D}_i g, \tag{3.1.10}$$

$$\Delta_h(\|x\|^\rho g) = \rho(2n+2\lambda_k+\rho)\|x\|^{\rho-2} g + \|x\|^\rho \Delta_h g. \tag{3.1.11}$$

The recurrence relation follows immediately from (3.1.10), which shows, by induction, that p_α is homogeneous of degree n. Using (3.1.11), a quick computation shows that $\Delta_h p_\alpha(x) = 0$, so that $p_\alpha \in \mathscr{H}_n^d(h_\kappa^2)$.

However, for a generic reflection group, it is not clear how to evaluate the norm of p_α, that is, an explicit formula for the norm of p_α is not known. As a consequence, we cannot provide an explicit orthonormal basis from $\{p_\alpha\}$. In fact, no orthonormal basis for $\mathscr{H}_n^d(h_\kappa^2)$ is explicitly known beyond the case of the group $\mathbb{Z}_2^d$. For $\mathbb{Z}_2^d$, the norm of p_α can be evaluated so that an orthonormal basis can be derived via the Gram–Schmidt process. In fact, for $\mathbb{Z}_2^d$, an orthonormal basis can be explicitly given in spherical coordinates, as h_κ^2 in this case is a simple product.

Let proj_n^κ denote the orthogonal projection operator

$$\mathrm{proj}_n^\kappa : L^2(\mathbb{S}^{d-1}; h_\kappa^2) \mapsto \mathscr{H}_n^d(h_\kappa^2).$$

By the orthogonal decomposition of homogeneous polynomials, $p \in \mathscr{P}_n^d$ can be written as $p(x) = p_n + \|x\|^2 q_n$, where $p_n \in \mathscr{H}_n^d(h_\kappa^2)$ and $q_n \in \mathscr{P}_{n-2}^d$; we have, by definition, that $p_n = \mathrm{proj}_n^\kappa p$. In this regard, by (3.1.9), the polynomial p_α in (3.1.8) is the orthogonal projection of x^α,

$$p_\alpha(x) = \mathrm{proj}_n^\kappa q_\alpha(x), \qquad q_\alpha(x) = x^\alpha \text{ with } |\alpha| = n. \tag{3.1.12}$$

Proposition 3.1.7. *Let $p \in \mathscr{P}_n^d$. Then*

$$\mathrm{proj}_n^\kappa p = \sum_{j=0}^{\lfloor n/2 \rfloor} \frac{1}{4^j j! (1-n-\lambda_\kappa)_j} \|x\|^{2j} \Delta_h^j p. \tag{3.1.13}$$

Proof. It suffices to prove equation (3.1.13) for $p = q_\alpha$. Since $\mathrm{proj}_n^\kappa q_\alpha = p_\alpha$, we need to show that p_α equals the right-hand side of (3.1.13) with $p = q_\alpha$. This can be established by induction on the degree $|\alpha|$ of q_α. Indeed, the case $|\alpha| = 1$ is obvious. Suppose equation (3.1.13) has been proved for all α such that $|\alpha| = n$, which gives

$$\mathscr{D}^\alpha\left\{\|x\|^{-2\lambda}\right\} = (-1)^n 2^n (\lambda)_n \|x\|^{-2\lambda-2n} \sum_{j=0}^{[n/2]} \frac{1}{4^j j! (-\lambda-n+1)_j} \|x\|^{2j} \Delta_h^j q_\alpha(x),$$

where $\lambda = \lambda_\kappa$. Applying here $\mathscr{D}_i$ and using the first identity in (3.1.10) with $g = \Delta_h^j\{x^\alpha\}$, we conclude that

$$\begin{aligned} \mathscr{D}_i \mathscr{D}^\alpha\left\{\|x\|^{-2\lambda}\right\} &= (-1)^n 2^n (\lambda)_n (-2\lambda - 2n) \|x\|^{-2\lambda-2n-2} \\ &\times \sum_{j=0}^{[(n+1)/2]} \frac{1}{4^j j! (-\lambda-n)_j} \|x\|^{2j} \left[x_i \Delta_h^j\{x^\alpha\} + 2j \Delta_h^{j-1} \mathscr{D}_i\{x^\alpha\}\right], \end{aligned} \tag{3.1.14}$$

since $\mathscr{D}_i$ commutes with Δ_h. From (3.1.10) and (3.1.11) it is easy to see, by induction on j, that

$$\Delta_h^j\{x_i f(x)\} = x_i \Delta_h^j f(x) + 2j \mathscr{D}_i \Delta_h^{j-1} f(x), \qquad j = 1, 2, 3, \ldots.$$

Thus, by the definition of p_α and (3.1.12), the left-hand side of (3.1.14) is a constant multiple of the projection of $x_i q_\alpha(x)$, and the right-hand side is a constant multiple of the right-hand side of (3.1.13) with $p(x) = x_i q_\alpha(x)$, which completes the induction. □

There is another inner product on the space of homogeneous polynomials $\mathscr{P}_n^d$ that will be useful in our study below.

Theorem 3.1.8. *For $p,q \in \mathscr{P}_n^d$,*

$$\langle p,q\rangle_{\mathscr{D}} = E_n(\mathscr{D}^{(x)}, \mathscr{D}^{(y)})p(x)q(y) = \langle q,p\rangle_{\mathscr{D}}.$$

Proof. By Proposition 2.3.8, $p(x) = E_n(x, \mathscr{D}^{(y)})p(y)$. The operators $\mathscr{D}^{(x)}$ and $\mathscr{D}^{(y)}$ commute and thus

$$\langle p,q\rangle_{\mathscr{D}} = E_n(\mathscr{D}^{(x)}, \mathscr{D}^{(y)})p(y)q(x) = E_n(\mathscr{D}^{(y)}, \mathscr{D}^{(x)})p(y)q(x).$$

The last expression equals $\langle q,p\rangle_{\mathscr{D}}$. □

The pairing $\langle\cdot,\cdot\rangle_{\mathscr{D}}$ is related to $\langle\cdot,\cdot\rangle_\kappa$ when $p \in \mathscr{P}_n^d$ and $q \in \mathscr{H}_n^d(h_\kappa^2)$.

Theorem 3.1.9. *If $p \in \mathscr{P}_n^d$ and $q \in \mathscr{H}_n^d(h_\kappa^2)$, then*

$$\langle p,q\rangle_{\mathscr{D}} = 2^n(\lambda_\kappa + 1)_n \langle p,q\rangle_\kappa. \tag{3.1.15}$$

Proof. Since $p(\mathscr{D})q(x)$ is a constant, it follows from Theorem 2.2.4 that

$$\begin{aligned}\langle p,q\rangle_h &= c_h \int_{\mathbb{R}^d} p(\mathscr{D})q(x)h_\kappa^2(x)e^{-\|x\|^2/2}dx \\ &= c_h \int_{\mathbb{R}^d} q(x)\big(p(\mathscr{D}^*)1\big)h_\kappa^2(x)e^{-\|x\|^2/2}dx.\end{aligned}$$

Repeatedly applying the formula for the adjoint operator $\mathscr{D}_i^* g(x) = x_i g(x) - \mathscr{D}_i g(x)$, and using the fact that the degree of $\mathscr{D}_i g$ is lower than that of g, we see that $p(\mathscr{D}^*)1 = p(x) + s(x)$ with a polynomial s of degree less than n. Since $q \in \mathscr{H}_n^d(h_\kappa^2)$, it follows from the spherical polar integral that $\int_{\mathbb{R}^d} q(x)s(x)h_\kappa^2(x)e^{-\|x\|^2/2}dx = 0$. Consequently, we conclude that

$$\langle p,q\rangle_h = c_h \int_{\mathbb{R}^d} q(x)p(x)h_\kappa^2(x)e^{-\|x\|^2/2}dx.$$

Now, (3.1.15) follows from Proposition 2.1.1. □

In the case where both $p,q \in \mathscr{P}_n^d$, the inner product $\langle p,q\rangle_{\mathscr{D}}$ also has an integral expression:

Theorem 3.1.10. *For $p,q \in \mathscr{P}_n^d$,*

$$\langle p,q\rangle_{\mathscr{D}} = c_h \int_{\mathbb{R}^d} \left(e^{-\Delta_h/2}p(x)\right)\left(e^{-\Delta_h/2}q(x)\right)h_\kappa^2(x)e^{-\|x\|^2/2}dx. \tag{3.1.16}$$

Proof. First of all, decomposing $p \in \mathscr{P}_n^d$ as $p(x) = \sum_{0 \le j \le n/2} \|x\|^{2j} p_{n-2j}$, where $p_{n-2j} \in \mathscr{H}_{n-2j}^d(h_\kappa^2)$, and decomposing q similarly, by the definition of Δ_h and $\langle p_{n-2j}, q_{n-2j} \rangle_h$, we conclude that

$$\langle p,q\rangle_h = \sum_{j=0}^{\lfloor n/2\rfloor}\sum_{i=0}^{\lfloor n/2\rfloor} \langle p_{n-2j}, q_{n-2j}\rangle_h = \sum_{j=0}^{\lfloor n/2\rfloor}\sum_{i=0}^{\lfloor n/2\rfloor} \Delta_h^i p_{n-2i}(\mathscr{D})\big(\|x\|^{2j} q_{n-2j}(x)\big).$$

By the identity (3.1.11),

$$\Delta_h^i\big(\|x\|^{2j} q_{n-2j}(x)\big) = 4^i(-j)_i(-n-\lambda+j)_i \|x\|^{2j-2i} q_{n-2j}(x),$$

which is zero if $i > j$. If $i < j$, then $\langle \|x\|^{2j} q_{n-2j}(x), \|x\|^{2i} p_{n-2i}(x)\rangle_h = 0$ by the same argument and the fact that $\langle p,q\rangle = \langle q,p\rangle$. Hence, the only remaining terms are those with $j = i$, which are given by $4^j j!(-n-\lambda+j)_j p_{n-2j}(\mathscr{D}) q_{n-2j}(x)$. Therefore,

$$\langle p,q\rangle_h = \sum_{j=0}^{\lfloor n/2\rfloor} 4^j j!(n-2j+\lambda_\kappa+1)_j \langle p_{n-2j}, q_{n-2j}\rangle_h.$$

Thus, we only need to prove (3.1.16) for polynomials of the form $p(x) = \|x\|^{2j} p_m(x)$ and $q(x) = \|x\|^{2j} q_m(x)$ with $p_m, q_m \in \mathscr{H}_{n-2j}^d(h_\kappa^2)$. For such p, q, by (3.1.11) and (3.1.15),

$$\begin{aligned}\langle p,q\rangle_{\mathscr{D}} &= 4^j j!(m+\lambda_\kappa+1)_j \langle p_m, q_m\rangle_{\mathscr{D}} \\ &= 4^j j!(m+\lambda_\kappa+1)_j 2^m(\lambda_\kappa+1)_j \langle p_m, q_m\rangle_\kappa \\ &= 2^{m+2j} j!(\lambda_\kappa+1)_{m+j} \langle p_m, q_m\rangle_\kappa.\end{aligned}$$

On the other hand, a straightforward computation using (3.1.11) shows that

$$e^{-\Delta_h/2}\left[\|x\|^{2j} p_m(x)\right] = (-1)^j j! 2^j L_j^{n+\lambda_\kappa}(\|x\|^2/2) p_m(x), \tag{3.1.17}$$

where L_j^α is the standard Laguerre polynomial. As a consequence, the right-hand side of the stated formula becomes, using spherical polar coordinates,

$$\begin{aligned}&\int_{\mathbb{R}^d} [L_j^{m+\lambda_\kappa}(\|x\|^2/2)]^2 p_m(x) q_m(x) h_\kappa^2(x) e^{-\|x\|^2/2} dx \\ &\quad = 2^{m+\lambda_\kappa} \frac{\Gamma(m+j+\lambda_\kappa+1)}{j!} \int_{\mathbb{S}^{d-1}} p_m(\xi) q_m(\xi) h_\kappa^2(\xi) d\sigma.\end{aligned}$$

Putting these together we get (3.1.16). □

3.2 Projection operator and intertwining operator

Let $\{Y_{\nu,n} : 1 \le \nu \le a_n^d\}$, $a_n^d := \dim \mathscr{H}_n^d(h_\kappa^2)$, be an orthonormal basis of $\mathscr{H}_n^d(h_\kappa^2)$. For $f \in L^2(\mathbb{S}^{d-1}, h_\kappa^2)$, the usual Hilbert space theory shows that f can be expanded in spherical h-harmonics as

$$f = \sum_{n=0}^{\infty} \sum_{\nu=0}^{a_n^d} \widehat{f}_{\nu,n} Y_{\nu,n}, \qquad \widehat{f}_{\nu,n} := \langle f, Y_{\nu,n}\rangle_\kappa,$$

where the convergence holds in $L^2(\mathbb{S}^{d-1}; h_\kappa^2)$ norm. In terms of the orthogonal projection operator from $L^2(\mathbb{S}^{d-1}, h_\kappa^2)$ onto $\mathscr{H}_n^d(h_\kappa^2)$,

$$\operatorname{proj}_n^\kappa : L^2(\mathbb{S}^{d-1}; h_\kappa^2) \mapsto \mathscr{H}_n^d(h_\kappa^2),$$

the expansion in h-harmonics can be rewritten as

$$f = \sum_{n=0}^{\infty} \operatorname{proj}_n^\kappa f.$$

In particular, the projection operator can be expressed as an integral,

$$\operatorname{proj}_n^\kappa f(x) = \frac{1}{\omega_d^\kappa} \int_{\mathbb{S}^{d-1}} f(y) Z_n^\kappa(x,y) h_\kappa^2(y) d\sigma(y), \quad x \in \mathbb{S}^{d-1}, \tag{3.2.1}$$

where $Z_n^\kappa(\cdot,\cdot)$ is the kernel function defined by

$$Z_n^\kappa(x,y) = \sum_{\nu=0}^{a_n^d} Y_{\nu,n}(x) Y_{\nu,n}(y), \quad x, y \in \mathbb{S}^{d-1}.$$

This kernel, however, is independent of the choice of particular basis of $\mathscr{H}_n^d(h_\kappa^2)$. Indeed, it is the reproducing kernel of $\mathscr{H}_n^d(h_\kappa^2)$, i.e.,

$$\frac{1}{\omega_d} \int_{\mathbb{S}^{d-1}} Z_n^\kappa(x,y) p(y) h_\kappa^2(y) d\sigma(y) = p(x), \quad \forall p \in \mathscr{H}_n^d(h_\kappa^2), \quad x \in \mathbb{S}^{d-1}. \tag{3.2.2}$$

In terms of the intertwining operator V_κ, the reproducing kernel Z_n^κ has a concise expression given in terms of the Gegenbauer polynomial C_n^λ:

Theorem 3.2.1. *Let $\lambda_\kappa = \gamma_k + \frac{d-2}{2}$. For $\|y\| \le \|x\| = 1$,*

$$Z_n^\kappa(x,y) = \|y\|^n \frac{n+\lambda_\kappa}{\lambda_\kappa} V_\kappa \left[C_n^{\lambda_\kappa} \left(\left\langle \cdot, \frac{y}{\|y\|} \right\rangle \right) \right](x). \tag{3.2.3}$$

Proof. From Proposition 2.3.8 (ii), it follows that for every $p \in \mathscr{H}_n^d(h_\kappa^2)$,

$$\begin{aligned} p(x) = \langle E_n(x,\cdot), p \rangle_{\mathscr{D}} &= \langle \operatorname{proj}_n^\kappa(E_n(x,\cdot)), p \rangle_{\mathscr{D}} \\ &= 2^n (\lambda_k + 1)_n \langle \operatorname{proj}_n^\kappa(E_n(x,\cdot)), p \rangle_\kappa. \end{aligned}$$

Since $Z_n^\kappa(\cdot,\cdot)$ is uniquely determined by the reproducing property, this shows that

$$Z_n^\kappa(x,\cdot) = 2^n (\lambda_k + 1)_n \operatorname{proj}_n^\kappa(E_n(x,\cdot)).$$

Using the intertwining property of V_κ and the definition of $E_n(\cdot,\cdot)$, it follows from (3.1.13) that

$$Z_n^\kappa(x,y) = \sum_{0 \le j \le n/2} \frac{(\lambda_\kappa + 1)_n 2^{n-2j}}{(1-n-\lambda_\kappa)_j j!} \|x\|^{2j} \|y\|^{2j} E_{n-2j}(x,y).$$

When $\|x\| = 1$, we can write the right-hand side as $V_\kappa(G_n(\langle \cdot, y/\|y\| \rangle))(x)$, where G_n is a hypergeometric function ${}_2F_1$, which turns out to be a constant multiple of the Gegenbauer polynomial. □

The identity (3.2.3) also indicates that in the theory of h-harmonics zonal functions, which depend only on $\langle x,y\rangle$, should be replaced by functions of the form $V_\kappa[f(\langle\cdot,y\rangle)](x)$. Indeed, we have an analogue of the Funk–Hecke formula.

Theorem 3.2.2. *Let f be a continuous function on $[-1,1]$. Then for any $Y_n^h \in \mathscr{H}_n^d(h_\kappa^2)$,*

$$\frac{1}{\omega_d^\kappa}\int_{\mathbb{S}^{d-1}} V_\kappa[f(\langle x,\cdot\rangle)](y)Y_n^h(y)h_\kappa^2(y)d\sigma(y) = \Lambda_n(f)Y_n^h(x), \quad x\in\mathbb{S}^{d-1}, \tag{3.2.4}$$

where $\Lambda_n(f)$ is a constant defined by

$$\Lambda_n(f) = c_{\lambda_\kappa}\int_{-1}^1 f(t)\frac{C_n^{\lambda_\kappa}(t)}{C_n^{\lambda_\kappa}(1)}(1-t^2)^{\lambda_\kappa-\frac12}dt,$$

and where $c_\lambda = \Gamma(\lambda+1)/\sqrt{\pi}\Gamma(\lambda+1/2)$ and $\Lambda_0(1)=1$.

Proof. If f is a polynomial of degree m, then we can expand f in terms of the Gegenbauer polynomials

$$f(t) = \sum_{k=0}^m \Lambda_k \frac{k+\lambda_\kappa}{\lambda_\kappa} C_k^{\lambda_\kappa}(t),$$

where Λ_κ are determined by the orthogonality of Gegenbauer polynomials,

$$\Lambda_k = \frac{c_{\lambda_\kappa}}{C_k^{\lambda_\kappa}(1)}\int_{-1}^1 f(t)C_k^{\lambda_\kappa}(t)(1-t^2)^{\lambda_\kappa-\frac12}dt,$$

and $c_\lambda^{-1} = \int_{-1}^1 (1-t^2)^{\lambda-\frac12}dt$. Using (3.2.3) and the reproducing property of $Z_n^\kappa(x,y)$ it follows that, for $n\le m$,

$$\frac{1}{\omega_d^\kappa}\int_{\mathbb{S}^{d-1}} V_\kappa[f(\langle x,\cdot\rangle)](y)Y_n^h(y)h_\kappa^2(y)d\sigma(y) = \Lambda_n Y_n^h(x), \quad x\in\mathbb{S}^{d-1}.$$

Since $\Lambda_n/\omega_d^\kappa = \Lambda_n(f)$ by definition, we have established the Funk–Hecke formula (3.2.4) for polynomials, and hence, by the Weierstrass theorem, for continuous functions. □

Theorem 3.2.3. *Let $f:\mathbb{B}^d\to\mathbb{R}$ be a continuous function. Then*

$$\frac{1}{\omega_d^\kappa}\int_{\mathbb{S}^{d-1}} V_k f(y)h_\kappa^2(y)d\sigma(y) = a_\kappa\int_{\mathbb{B}^d} f(x)(1-\|x\|^2)^{|\kappa|-1}dx. \tag{3.2.5}$$

In particular, if $f(y)=g(\langle x,y\rangle)$ with $g:\mathbb{R}\to\mathbb{R}$, then

$$\frac{1}{\omega_d^\kappa}\int_{\mathbb{S}^{d-1}} V_k[g(\langle x,\cdot\rangle)](y)h_\kappa^2(y)d\sigma(y) = c_{\lambda_\kappa}\int_{-1}^1 g(\|x\|t)(1-t^2)^{\lambda_\kappa-\frac12}dt. \tag{3.2.6}$$

Proof. Applying the Funk–Hecke formula (3.2.4) with $n=0$ to the function $\xi\mapsto g(\|x\|\xi)$, $\xi\in\mathbb{S}^{d-1}$, gives (3.2.6). It is known that every polynomial f can be written as a linear sum of $p_j(\langle x,\xi_j\rangle)$, where $p_j:[-1,1]\to\mathbb{R}$ and $\xi_j\in\mathbb{S}^{d-1}$, so that (3.2.5) follows from (3.2.6) for all polynomials. For general f we can then pass to the limit, since the right-hand side of (3.2.5) is clearly closed under limits. □

Remark 3.2.4. Using Theorem 3.2.3 and the positivity of V_κ, for any $g \in C(\mathbb{B}^d)$,

$$\|V_\kappa g\|_{L^1(h_\kappa^2;\mathbb{S}^{d-1})} \le b_\kappa \int_{\mathbb{B}^d} |g(x)|(1-\|x\|^2)^{|\kappa|-1} dx.$$

Since the right-hand side is a constant multiple of the norm of $L^1(W;\mathbb{B}^d)$, where $W(x) := (1-\|x\|^2)^{|\kappa|-1}$, this allows us to extend V_κ to a positive, bounded operator from $L^1(W;\mathbb{B}^d)$ to $L^1(h_\kappa^2;\mathbb{S}^{d-1})$, so that (3.2.5) holds for all $g \in L^1(W;\mathbb{B}^d)$.

For a generic reflection group, not very much is known on specifics of the intertwining operator V_κ. Property (3.2.5) of V_κ is highly non-trivial, as can be seen in the special case of $\mathbb{Z}_2^d$, where V_κ is given explicitly by formula (2.3.2), and it is highly useful, as the development below will show.

3.3 Convolution operators and orthogonal expansions

The expression of $Z_n^\kappa(\cdot,\cdot)$ at (3.2.3) suggests the following definition of convolution on the sphere.

Definition 3.3.1. For $f \in L^1(\mathbb{S}^{d-1};h_\kappa^2)$ and $g \in L^1(w_{\lambda_\kappa},[-1,1])$,

$$(f *_\kappa g)(x) := \frac{1}{\omega_d^\kappa} \int_{\mathbb{S}^{d-1}} f(y) V_\kappa[g(\langle\cdot,y\rangle)](x) h_\kappa^2(y) d\sigma(y). \tag{3.3.1}$$

Denote the norm of the space $L^p(w_\lambda;[-1,1])$ by $\|\cdot\|_{\lambda,p}$, and the norm of the space $L^p(h_\kappa,\mathbb{S}^{d-1})$ by $\|\cdot\|_{\kappa,p}$ for $1 \le p < \infty$ and by $C(\mathbb{S}^{d-1})$ for $p=\infty$. The convolution $*_\kappa$ satisfies Young's inequality:

Theorem 3.3.2. *Let $p,q,r \ge 1$ and $p^{-1} = r^{-1} + q^{-1} - 1$. For $f \in L^q(\mathbb{S}^{d-1};h_\kappa^2)$ and $g \in L^r(w_{\lambda_\kappa};[-1,1])$,*

$$\|f *_\kappa g\|_{\kappa,p} \le \|f\|_{\kappa,q} \|g\|_{\lambda_\kappa,r}. \tag{3.3.2}$$

In particular, for $1 \le p \le \infty$,

$$\|f * g\|_{\kappa,p} \le \|f\|_{\kappa,p} \|g\|_{\lambda_\kappa,1} \quad \textit{and} \quad \|f * g\|_{\kappa,p} \le \|f\|_{\kappa,1} \|g\|_{\lambda_\kappa,p}. \tag{3.3.3}$$

Proof. The standard proof of Young's inequality applies in this setting. By Minkowski's inequality, it suffices to show that

$$\|G(x,\cdot)\|_{\kappa,r} \le \|g\|_{\lambda_\kappa,r}, \qquad \text{where} \quad G(x,y) = V_\kappa[g(\langle x,\cdot\rangle)](y).$$

The proof uses the integral relation (3.2.6). Indeed, the positivity of V_κ implies $|V_\kappa g| \le V_\kappa[|g|]$, so that $\|G(x,\cdot)\|_{\kappa,\infty} \le \|g\|_{\lambda_\kappa,\infty}$ and we deduce by (3.2.6) that

$$\|G(x,\cdot)\|_{\kappa,1} \le \frac{1}{\omega_d^\kappa} \int_{\mathbb{S}^{d-1}} V_\kappa[|g(\langle x,\cdot\rangle)|](y) h_\kappa^2(y) d\sigma = c_{\lambda_\kappa} \int_{-1}^1 |g(t)| w_\lambda(t) dt = \|g\|_{\lambda_\kappa,1}.$$

The log-convexity of the L^p-norm implies then $\|G(x,\cdot)\|_{\kappa,r} \le \|g\|_{\lambda_\kappa,r}$. □

By (3.2.1) and (3.2.3), the projection proj_n^κ is a convolution operator

$$\mathrm{proj}_n^\kappa f = f *_\kappa Z_n^\kappa, \qquad Z_n^\kappa(t) := \frac{n+\lambda_\kappa}{\lambda_\kappa} C_n^{\lambda_\kappa}(t). \tag{3.3.4}$$

The following theorem justifies calling $*_\kappa$ a convolution:

Theorem 3.3.3. *For $f \in L^1(\mathbb{S}^{d-1}; h_\kappa^2)$ and $g \in L^1(w_{\lambda_\kappa}; [-1,1])$,*

$$\mathrm{proj}_n^\kappa (f *_\kappa g) = \widehat{g}_n^{\lambda_\kappa} \, \mathrm{proj}_n^\kappa f, \quad n = 0,1,2\ldots, \tag{3.3.5}$$

where $\widehat{g}_n^{\lambda_\kappa}$ is the Fourier coefficient of g in the Gegenbauer polynomial,

$$\widehat{g}_n^{\lambda_\kappa} = c_{\lambda_\kappa} \int_{-1}^1 g(t) \frac{C_n^{\lambda_\kappa}(t)}{C_n^{\lambda_\kappa}(1)} (1-t^2)^{\lambda_\kappa - \frac12} dt.$$

Proof. By (3.2.1) and the Funk–Hecke formula in Theorem 3.2.2,

$$\begin{aligned}
\mathrm{proj}_n^\kappa (f *_\kappa g)(x) &= \frac{1}{\omega_d^\kappa} \int_{\mathbb{S}^{d-1}} (f*g)(\xi) Z_n^\kappa(x,\xi) h_\kappa^2(\xi) d\sigma(\xi) \\
&= \frac{1}{\omega_d^\kappa} \int_{\mathbb{S}^{d-1}} f(y) \left(\frac{1}{\omega_d^\kappa} \int_{\mathbb{S}^{d-1}} g(\langle \xi, y \rangle) Z_n^\kappa(x,\xi) h_\kappa^2(\xi) d\sigma(\xi) \right) h_\kappa^2(y) d\sigma(y) \\
&= \widehat{g}_n^{\lambda_\kappa} \frac{1}{\omega_d^\kappa} \int_{\mathbb{S}^{d-1}} f(y) Z_n^\kappa(x,y) h_\kappa^2(y) d\sigma(y) = \widehat{g}_n^{\lambda_\kappa} \, \mathrm{proj}_n^\kappa f(x),
\end{aligned}$$

which is what we needed to prove. □

Since the convergence of the h-harmonic series does not go beyond L^2 norm in general, it is necessary to consider summability methods. We consider the Cesàro means of the h-harmonic series. First, we give the definition of the Cesàro means for a sequence of complex numbers.

Definition 3.3.4. The Cesàro (C,δ)-means of a given sequence $\{a_n\}_{n=0}^\infty$ of complex numbers are defined by

$$s_n^\delta := \sum_{j=0}^n \frac{A_{n-j}^\delta}{A_n^\delta} a_j, \quad n = 0,1,\ldots, \tag{3.3.6}$$

where the coefficients A_j^δ are defined by

$$(1-t)^{-1-\delta} = \sum_{n=0}^\infty A_n^\delta t^n, \quad t \in (-1,1).$$

For convenience, we also define $A_j^\delta = 0$ for $j < 0$. The following useful properties follow easily from the definition:

$$A_j^\delta - A_{j-1}^\delta = A_j^{\delta-1}, \qquad \sum_{j=0}^n A_j^\delta = A_n^{\delta+1}, \quad \sum_{j=0}^n A_{n-j}^\delta A_j^\alpha = A_n^{\alpha+\delta+1},$$
$$|A_j^\delta| \sim (j+1)^\delta, \quad \text{whenever } j+\delta+1 > 0.$$

Let us denote by $S_n^\delta\left(h_\kappa^2;f\right)$ the Cesàro means of the h-harmonic series, that is,

$$S_n^\delta\left(h_\kappa^2;f\right) := \frac{1}{A_n^\delta}\sum_{j=0}^{n} A_{n-j}^\delta \operatorname{proj}_j^\kappa f, \tag{3.3.7}$$

where $S_n^0(h_\kappa^2;f)$ is the n-th partial sum. By (3.3.4), the Cesàro means are convolution operators,

$$S_n^\delta\left(h_\kappa^2;f\right) = f *_\kappa K_n^\delta\left(h_\kappa^2\right),$$

where the kernel is defined by

$$K_n^\delta\left(h_\kappa^2;t\right) := \frac{1}{A_n^\delta}\sum_{k=0}^{n} A_{n-k}^\delta \frac{k+\lambda_\kappa}{\lambda_\kappa} C_k^{\lambda_\kappa}(t) = k_n^\delta\left(w_{\lambda_\kappa};1,t\right), \tag{3.3.8}$$

in which $k_n^\delta(w_{\lambda_\kappa};\cdot,\cdot)$ is the kernel of the (C,δ)-means of the Fourier orthogonal series in the Gegenbauer polynomials.

Theorem 3.3.5. *The Cesàro means of the spherical h-harmonic series satisfy:*

1. *if $\delta \ge 2\lambda_k + 1$, then $S_n^\delta(h_\kappa^2)$ is a nonnegative operator;*
2. *if $\delta > \lambda_\kappa$, then*

$$\sup_{n\ge 0} \|S_n^\delta(h_\kappa^2;g)\|_{\kappa,p} \le c\|g\|_{\kappa,p}, \quad 1 \le p \le \infty. \tag{3.3.9}$$

In particular, $S_n^\delta(h_\kappa^2;f)$ converges to f in $L^p(h_\kappa^2;\mathbb{S}^{d-1})$ for $1 \le p \le \infty$.

Proof. By (3.3.8), the non-negativity of $S_n^\delta(h_\kappa^2)$ follows from that of the Gegenbauer kernel $k_n^\delta(w_\lambda;1,t)$, which is a classical result. In order to prove the convergence, it is sufficient to show that $\|S_n^\delta(h_k^2)\|_{\kappa,p}$ is bounded. By Young's inequality (3.3.3),

$$\|S_n^\delta(h_\kappa^2,f)\|_{\kappa,p} \le \|f\|_{\kappa,p}\|k_n^\delta(w_{\lambda_\kappa})\|_{\lambda_\kappa,1} \le c\|f\|_{\kappa,p},$$

whenever $\delta > \lambda_k$, where the last inequality follows from a classical result on the (C,δ) summability of the Gegenbauer series. □

A careful examination of the proof of convergence shows that the essential ingredients are (3.2.6) and the positivity of V_κ (namely, Theorem 2.3.4). Equality (3.2.6) removes V_κ and allows us to reduce the problem to the Gegenbauer series. This is similar to the result for ordinary spherical harmonics. However, for the ordinary spherical harmonics, this leads to the sharp condition $\delta > (d-2)/2$, while for the h-harmonic series, $\delta > \lambda_\kappa$ is not sharp in general. In fact, taking the average of V_κ by (3.2.6) erases the information on the reflection group.

For a generic reflection group, we know very little about V_κ. In the case of $\mathbb{Z}_2^d$, however, V_κ is given by the explicit formula in (2.3.2), which allows us to obtain much deeper results on the convergence of the Cesàro means. These studies require sharp pointwise estimates of the kernel functions and, sometimes, asymptotics of the kernel functions, which in turn require long estimations and detailed analysis. We shall state the results without proof.

Recall that, for the h_κ defined in (2.1.1) associated with $\mathbb{Z}_2^d$,

$$\lambda_\kappa = |\kappa| + \frac{d-2}{2}.$$

Theorem 3.3.6. *Let h_κ be as in* (2.1.1). *Let $\delta > -1$ and define*

$$\sigma_\kappa := \lambda_k - \min_{1\le i\le d} \kappa_i = |\kappa| + \frac{d-2}{2} - \min_{1\le i\le d} \kappa_i.$$

Then for $p = 1$ and $p = \infty$,

$$\|\operatorname{proj}_n(h_\kappa^2)\|_{\kappa,p} \sim n^{\sigma_\kappa} \quad \textit{and} \quad \|S_n^\delta(h_\kappa^2)\|_{\kappa,p} \sim \begin{cases} 1, & \delta > \sigma_\kappa \\ \log n, & \delta = \sigma_\kappa \\ n^{-\delta+\sigma_\kappa}, & -1 < \delta < \sigma_\kappa. \end{cases}$$

In particular, the (C,δ)-means $S_n^\delta(h_\kappa^2; f)$ converge to f in $L^1(h_\kappa^2; \mathbb{S}^{d-1})$ or in $C(\mathbb{S}^{d-1})$ if and only if $\delta > \sigma_\kappa$.

The zero set of the weight function $h_\kappa(x)$ serves as boundary on the sphere, away from which we have better convergence behavior. Indeed, for h_κ in (2.1.1), the zero set is the collection of great circles defined by the intersection of $\mathbb{S}^{d-1}$ with the coordinate planes. Let us define

$$\mathbb{S}_{\text{int}}^{d-1} := \mathbb{S}^{d-1} \setminus \bigcup_{i=1}^{d} \{x \in \mathbb{S}^{d-1} : x_i = 0\},$$

which is the interior region bounded by these great circles on $\mathbb{S}^{d-1}$.

Theorem 3.3.7. *Let h_κ be as in* (2.1.1). *Let f be continuous on $\mathbb{S}^{d-1}$. If $\delta > \frac{d-2}{2}$, then $S_n^\delta(h_\kappa^2; f)$ converges to f for every $x \in \mathbb{S}_{\text{int}}^{d-1}$ and the convergence is uniform over each compact subset of $\mathbb{S}_{\text{int}}^{d-1}$.*

By the Riesz interpolation theorem, Theorem 3.3.6 also implies that $S_n^\delta(h_\kappa^2; f)$ converges to f in the L^p norm if $\delta > \sigma_\kappa$, for all $1 < p < \infty$. However, for each fixed p, this order is not sharp. The sharp results for L^p convergence are given in the following two theorems:

Theorem 3.3.8. *Suppose that $f \in L^p(\mathbb{S}^{d-1}; h_\kappa^2)$, $1 \le p \le \infty$, $|\frac{1}{p} - \frac{1}{2}| \ge \frac{1}{2\sigma_\kappa+2}$ and*

$$\delta > \delta_\kappa(p) := \max\{(2\sigma_\kappa + 1)|\tfrac{1}{p} - \tfrac{1}{2}| - \tfrac{1}{2}, 0\}. \tag{3.3.10}$$

Then $S_n^\delta(h_\kappa^2; f)$ converges to f in $L^p(h_\kappa^2; \mathbb{S}^{d-1})$ and

$$\sup_{n\in\mathbb{N}} \|S_n^\delta(h_\kappa^2; f)\|_{\kappa,p} \le c\|f\|_{\kappa,p}.$$

Theorem 3.3.9. *Assume $1 \le p \le \infty$ and $0 < \delta \le \delta_\kappa(p)$. Then there exists a function $f \in L^p(\mathbb{S}^{d-1}; h_\kappa^2)$ such that $S_n^\delta(h_\kappa^2; f)$ diverges in $L^p(\mathbb{S}^{d-1}; h_\kappa^2)$.*

The proofs of these two theorems are much more involved and require heavy machinery, such as the Fefferman–Stein inequality and Stein's analytic interpolation theorem.

3.4 Maximal functions

For $x \in \mathbb{S}^{d-1}$ and $0 \le \theta \le \pi$, we define

$$\mathtt{b}(x,\theta) := \{y \in \mathbb{B}^d : \langle x, y\rangle \ge \cos\theta\}.$$

Let χ_E denote the characteristic function of the set E.

Definition 3.4.1. For $f \in L^1(\mathbb{S}^{d-1}; h_\kappa^2)$, define the maximal function

$$\begin{aligned} \mathscr{M}_\kappa f(x) &= \sup_{0<\theta\le\pi} \frac{\int_{\mathbb{S}^{d-1}} |f(y)| V_\kappa[\chi_{\mathtt{b}(x,\theta)}](y) h_\kappa^2(y) d\sigma(y)}{\int_{\mathbb{S}^{d-1}} V_\kappa[\chi_{\mathtt{b}(x,\theta)}](y) h_\kappa^2(y) d\sigma(y)} \\ &= \sup_{0<\theta\le\pi} \frac{(|f| *_\kappa \chi_{[\cos\theta,1]})(x)}{c_{\lambda_\kappa} \int_0^\theta (\sin\phi)^{2\lambda_\kappa} d\phi}. \end{aligned} \tag{3.4.1}$$

The second expression in this definition uses the identity

$$\int_{\mathbb{S}^{d-1}} V_\kappa[\chi_{[\cos\theta,1]}(\langle x,\cdot\rangle)](y) h_\kappa^2(y) d\sigma(y) = \int_0^\theta (\sin\phi)^{2\lambda_\kappa} d\phi \sim \theta^{2\lambda_\kappa+1}, \tag{3.4.2}$$

coming from (3.2.6). This maximal function can be used to study the h-harmonic expansions, since we can often prove that $|(f *_\kappa g)(x)| \le c\mathscr{M}_\kappa f(x)$. We will show that it satisfies the usual property of maximal functions, that is, it is of strong type (p,p) for $1 < p < \infty$, and of weak type $(1,1)$. The proof of this last result relies on a general result about the following semi-groups of operators (see [49, p. 2] for more details):

Definition 3.4.2. Let (X,μ) be a measure space with a positive measure μ. A family of operators $\{T^t\}_{t\ge 0}$ is said to form a symmetric diffusion semi-group if

$$T^{t_1}T^{t_2} = T^{t_1+t_2}, \qquad T^0 = \mathrm{id},$$

and

(i) T^t are contractions on $L^p(X,\mu)$, i.e., $\|T^t f\|_p \le \|f\|_p$, $1 \le p \le \infty$;

(ii) T^t are symmetric, i.e., each T^t is self-adjoint on $L^2(X, d\mu)$;

(iii) T^t are positivity preserving, i.e., $T^t f \ge 0$ if $f \ge 0$;

(iv) $T^t f_0 = f_0$ if $f_0(x) = 1$.

The result that we shall need is given in [49, p. 48] and it is a special case of the Hopf–Dunford–Schwartz ergodic theorem.

Theorem 3.4.3. *Suppose that $\{T^t\}_{t\ge 0}$ is a symmetric diffusion semi-group on a positive measure space (X,μ). Then the function*

$$Mf(x) = \sup_{s\ge 0}\left(\frac{1}{s}\int_0^s T^t f(x)dt\right)$$

satisfies the inequalities

(a) $\|Mf\|_p \le c_p\|f\|_p$ *for each p with $1<p\le\infty$;*

(b) $\mu(\{x\in X : Mf(x)>\alpha\}) \le (c/\alpha)\|f\|_1$ *for each $\alpha>0$ and $f\in L^1(X,\mu)$, where c is independent of f and α.*

Our semi-group of operators is defined in terms of the Poisson integrals:

Definition 3.4.4. For $f\in L^1(\mathbb{S}^{d-1};h_\kappa^2)$, the Poisson integral of f is defined by

$$P_r^\kappa f(\xi) := \frac{1}{\omega_d^\kappa}\int_{\mathbb{S}^{d-1}} f(y)P_r^\kappa(\xi,y)h_\kappa^2(y)d\sigma(y), \quad \xi\in\mathbb{S}^{d-1}, \tag{3.4.3}$$

where the kernel $P_r^\kappa(x,\cdot)$ is given by

$$P_r^\kappa(x,y) := V_\kappa\left[\frac{1-r^2}{(1-2r\langle\cdot,y\rangle+r^2)^{\lambda_\kappa+1}}\right](x), \tag{3.4.4}$$

for $0<r<1$.

Using the generating function of the Gegenbauer polynomials, the Poisson kernel is the generating function of the spherical h-harmonics, as seen in the first item of the following lemma, from which the other two items follow directly.

Lemma 3.4.5. *For $0<r<1$, the Poisson kernel satisfies the following properties:*

(1) *for $x,y\in\mathbb{S}^{d-1}$, $P_r^\kappa(x,y) = \sum_{n=0}^\infty r^n\frac{n+\lambda_\kappa}{\lambda_\kappa}V_\kappa\left[C_n^{\lambda_\kappa}(\langle x,\cdot\rangle)\right](y)$;*

(2) *$P_r^\kappa f = \sum_{n=0}^\infty r^n \operatorname{proj}_n^\kappa f$;*

(3) *$P_r^\kappa(x,y)\ge 0$ and $\frac{1}{\omega_d^\kappa}\int_{\mathbb{S}^{d-1}}P_r^\kappa(x,y)h_\kappa^2(y)d\sigma(y)=1$.*

Put $T^t = P_r^\kappa$ with $r=e^{-t}$. Using the above lemma, it is easy to see that T^t is a diffusion semi-group. We will need another semi-group, which is the discrete analog of the heat operator:

$$H_t^\kappa f := f *_\kappa q_t^\kappa, \qquad q_t^\kappa(s) := \sum_{n=0}^\infty e^{-n(n+2\lambda_\kappa)t}\frac{n+\lambda_\kappa}{\lambda_\kappa}C_n^{\lambda_\kappa}(s). \tag{3.4.5}$$

Lemma 3.4.6. *The family of operators $\{H_t^\kappa\}$ is a symmetric diffusion semi-group.*

Proof. The kernel q_t^κ is known to be nonnegative, from which it immediately follows that H_t^κ are positive and that $\|q_t^\kappa\|_{\lambda_\kappa,1} = 1$, by the orthogonality of the Gegenbauer polynomials. Hence, by Young's inequality, $\|H_t^\kappa f\|_{\kappa,p} \le \|f\|_{k,p}$. The other requirements in Definition 3.4.2 can be verified directly. □

Lemma 3.4.7. *The Poisson and the heat semi-groups are connected by*

$$P_{e^{-t}}^\kappa f(x) = \int_0^\infty \phi_t(s) H_s^\kappa f(x) ds, \tag{3.4.6}$$

where

$$\phi_t(s) := \frac{t}{2\sqrt{\pi}} s^{-3/2} e^{-\left(\frac{t}{2\sqrt{s}} - \lambda_\kappa \sqrt{s}\right)^2}.$$

Furthermore, if $f(x) \ge 0$ for all x, then

$$P_*^\kappa f(x) := \sup_{0<r<1} P_r^\kappa f(x) \le c \sup_{s>0} \frac{1}{s} \int_0^s H_u^\kappa f(x) du. \tag{3.4.7}$$

Consequently, $P_*^\kappa f$ is bounded on $L^p(h_\kappa^2; \mathbb{S}^{d-1})$ for $1 < p \le \infty$, and of weak type $(1,1)$.

Proof. Since $\{H_t^\kappa\}$ is a semi-group of operators, by Theorem 3.4.3, the maximal operator $\sup_{s>0}\left(\frac{1}{s}\int_0^s H_u^\kappa f(x) du\right)$ is bounded on $L^p(h_\kappa^2, \mathbb{S}^{d-1})$ for $1 < p \le \infty$ and of week type $(1,1)$. Therefore, it is sufficient to prove (3.4.6) and (3.4.7).

First we prove (3.4.6). We use the well-known identity ([49, p.46])

$$e^{-v} = \frac{1}{\sqrt{\pi}} \int_0^\infty \frac{e^{-u}}{\sqrt{u}} e^{-v^2/4u} du, \qquad v > 0, \tag{3.4.8}$$

with $v = (n + \lambda_\kappa)t$. Making the change of variable $s = t^2/4u$, we obtain

$$\begin{aligned} e^{-nt} &= e^{\lambda_\kappa t} \frac{1}{\sqrt{\pi}} \int_0^\infty \frac{e^{-u}}{\sqrt{u}} e^{-\frac{n(n+2\lambda_\kappa)t^2}{4u}} e^{-\frac{\lambda_\kappa^2 t^2}{4u}} du \\ &= \frac{t}{2\sqrt{\pi}} \int_0^\infty e^{-n(n+2\lambda_\kappa)s} s^{-3/2} e^{-\left(\frac{t}{2\sqrt{s}} - \lambda_\kappa\sqrt{s}\right)^2} ds \\ &= \int_0^\infty e^{-n(n+2\lambda_\kappa)s} \phi_t(s)\, ds. \end{aligned}$$

Multiplying by $\operatorname{proj}_n^\kappa f$ and summing up over n we obtain the integral relation (3.4.6).

For the proof of (3.4.7), we use (3.4.6) and integration by parts to obtain

$$\begin{aligned} P_{e^{-t}}^\kappa f(x) &= -\int_0^\infty \left(\int_0^s H_u^\kappa f(x) du\right) \phi_t'(s) ds \\ &\le \sup_{s>0} \left(\frac{1}{s}\int_0^s H_u^\kappa f(x) du\right) \int_0^\infty s|\phi_t'(s)| ds, \end{aligned}$$

where the derivative of $\phi_t'(s)$ is taken with respect to s. Furthermore, since $P_r^\kappa f = f *_k p_r^\kappa$ and $|p_r^\kappa(t)| \le c$ for $0 < r \le e^{-1}$, it follows that

$$\sup_{0<r\le e^{-1}} P_r^\kappa f(x) \le c\|f\|_{1,\kappa} = c \lim_{s\to\infty} \frac{1}{s}\int_0^s H_u^\kappa(|f|)(x)\,du.$$

Therefore, to finish the proof of (3.4.7), it suffices to show that $\sup_{0<t\le 1}\int_0^\infty s|\phi_t'(s)|ds$ is bounded by a constant. A quick computation shows that $\phi_t'(s) > 0$ if $s < \alpha_t$ and $\phi_t'(s) < 0$ if $s > \alpha_t$, where

$$\alpha_t := \frac{t^2}{3+\sqrt{9+4\lambda_\kappa^2 t^2}} \sim t^2, \qquad 0 \le t \le 1.$$

Since the integral of $\phi_t(s)$ over $[0,\infty)$ is 1 and $\phi_t(s) \ge 0$, integration by parts gives

$$\begin{aligned}\int_0^\infty s|\phi_t'(s)|ds &= 2\alpha_t\phi_t(\alpha_t) - \int_0^{\alpha_t}\phi_t(s)ds + \int_{\alpha_t}^\infty \phi_t(s)ds \\ &\le 2\alpha_t\phi_t(\alpha_t) + 1 = \frac{t}{\sqrt{\pi\alpha_t}} e^{-\frac{(t-2\lambda_\kappa\alpha_t)^2}{4\alpha_t}} + 1 \le c,\end{aligned}$$

as desired. □

We are now ready to prove the main result on the maximal function. To state the weak type inequality, we define, for any measurable subset E of $\mathbb{S}^{d-1}$, the measure with respect to h_κ^2 as

$$\operatorname{meas}_\kappa E := \int_E h_\kappa^2(y)d\sigma(y).$$

Theorem 3.4.8. *If $f \in L^1(\mathbb{S}^{d-1}; h_\kappa^2)$, then $\mathscr{M}_\kappa f$ satisfies*

$$\operatorname{meas}_\kappa\{x \in \mathbb{S}^{d-1} : \mathscr{M}_\kappa f(x) \ge \alpha\} \le c\frac{\|f\|_{\kappa,1}}{\alpha}, \quad \forall \alpha > 0. \tag{3.4.9}$$

Furthermore, if $f \in L^p(\mathbb{S}^{d-1}; h_\kappa^2)$ for $1 < p \le \infty$, then $\|\mathscr{M}_\kappa f\|_{\kappa,p} \le c\|f\|_{\kappa,p}$.

Proof. From the definition of p_r^κ in (3.4.4), if $1-r \sim \theta$, then

$$\begin{aligned}p_r^\kappa(\cos\theta) &= \frac{1-r^2}{\left((1-r)^2+4r\sin^2\frac{\theta}{2}\right)^{\lambda_\kappa+1}} \\ &\ge c\frac{1-r^2}{\left((1-r)^2+r\theta^2\right)^{\lambda_\kappa+1}} \ge c(1-r)^{-(2\lambda_\kappa+1)}.\end{aligned}$$

For $j \ge 0$ define $r_j := 1-2^{-j}\theta$ and set $B_j := \left\{y \in \mathbb{B}^d : 2^{-j-1}\theta \le d(x,y) \le 2^{-j}\theta\right\}$. The lower bound of p_r^κ proved above shows that

$$\chi_{B_j}(y) \le c(2^{-j}\theta)^{2\lambda_k+1} p_{r_j}^\kappa(\langle x,y\rangle),$$

which immediately implies that

$$\chi_{\mathsf{b}(x,\theta)}(y) \le \sum_{j=0}^{\infty} \chi_{B_j}(y) \le c\,\theta^{2\lambda_k+1} \sum_{j=0}^{\infty} 2^{-j(2\lambda_\kappa+1)} p_{r_j}^{\kappa}(\langle x,y\rangle).$$

Since V_κ is a positive linear operator, applying V_κ to the above inequality we get

$$\begin{aligned}
&\int_{S^{d-1}} |f(y)| V_\kappa \left[\chi_{\mathsf{b}(x,\theta)}\right](y) h_\kappa^2(y) d\sigma(y) \\
&\quad \le c\,\theta^{2\lambda_\kappa+1} \sum_{j=0}^{\infty} 2^{-j(2\lambda_\kappa+1)} \int_{S^{d-1}} |f(y)| V_\kappa \left[p_{r_j}(\langle x,y\rangle)\right](y) h_\kappa^2(y) d\sigma(y) \\
&\quad = c\,\theta^{2\lambda_\kappa+1} \sum_{j=0}^{\infty} 2^{-j(2\lambda_\kappa+1)} P_{r_j}^{\kappa}(|f|;x) \\
&\quad \le c\,\theta^{2\lambda_\kappa+1} \sup_{0<r<1} P_r^{\kappa}(|f|;x).
\end{aligned}$$

Dividing by $\theta^{2\lambda_\kappa+1}$ and using the fact that

$$\frac{1}{\omega_d^{\kappa}} \int_{\mathbb{S}^{d-1}} V_\kappa[\chi_{\mathsf{b}(x,\theta)}](y) h_\kappa^2(y) d\sigma(y) = c_{\lambda_\kappa} \int_0^{\theta} (\sin\phi)^{2\lambda_\kappa} d\phi \sim \theta^{2\lambda_\kappa+1},$$

we have proved that $\mathscr{M}_\kappa f(x) \le c P_*^{\kappa}|f|(x)$. The desired result now follows from Lemma 3.4.7. □

3.5 Convolution and maximal function

The ordinary convolution operator is often defined in terms of the translation operator. For $f \in L^2(\mathbb{S}^{d-1})$ and $0 \le \theta \le \pi$, the translation operator T_θ is defined by

$$T_\theta f(x) := \frac{1}{\omega_{d-1}(\sin\theta)^{d-1}} \int_{\langle x,y\rangle=\cos\theta} f(y) d\ell_{x,\theta}(y),$$

where $d\ell_{x,\theta}$ is the Lebesgue measure on the set $\{y \in \mathbb{S}^{d-1} : \langle x,y\rangle = \cos\theta\}$. For the integral with respect to the $h_\kappa^2 d\sigma$, we do not have an explicit extension of $T_\theta f$, but we can define an extension as a multiplier operator.

Definition 3.5.1. For $0 \le \theta \le \pi$, the generalized translation operator T_θ^κ is defined by

$$\operatorname{proj}_n^{\kappa}(T_\theta^{\kappa} f) = \frac{C_n^{\lambda_\kappa}(\cos\theta)}{C_n^{\lambda_\kappa}(1)} \operatorname{proj}_n^{\kappa} f, \quad n = 0, 1, \ldots. \tag{3.5.1}$$

Since a function in $L^1(\mathbb{S}^{d-1}; h_\kappa^2)$ is uniquely defined by its orthogonal projections on $\mathscr{H}_n^d(h_\kappa^2)$, the generalized translation operator is well defined.

Proposition 3.5.2. *The operator T_θ^κ is well defined for all $f \in L^1(\mathbb{S}^{d-1}; h_\kappa^2)$ and enjoys the following properties:*

(i) *for $f \in L^2(h_\kappa^2, \mathbb{S}^{d-1})$ and $g \in L^1(w_{\lambda_\kappa}, [-1,1])$,*

$$(f *_\kappa g)(x) = c_{\lambda_\kappa} \int_0^\pi T_\theta^\kappa f(x) g(\cos\theta)(\sin\theta)^{2\lambda_\kappa} d\theta; \tag{3.5.2}$$

(ii) *T_θ^κ preserves positivity, i.e., $T_\theta^\kappa f \ge 0$ if $f \ge 0$;*

(iii) *for $f \in L^p(h_\kappa^2, \mathbb{S}^{d-1})$ if $1 \le p < \infty$, or $f \in C(\mathbb{S}^{d-1})$ if $p = \infty$,*

$$\|T_\theta^\kappa f\|_{\kappa,p} \le \|f\|_{\kappa,p} \qquad \text{and} \qquad \lim_{\theta \to 0} \|T_\theta^\kappa f - f\|_{\kappa,p} = 0; \tag{3.5.3}$$

(iv) *$T_\theta^\kappa f(-x) = T_{\pi-\theta}^\kappa(x)$.*

Proof. The equation (3.3.5) immediately implies

$$\operatorname{proj}_n^\kappa (f *_\kappa g)(x) = c_{\lambda_\kappa} \int_0^\pi \operatorname{proj}_n^\kappa (T_\theta^\kappa f)(x) g(\cos\theta)(\sin\theta)^{2\lambda_\kappa} d\theta,$$

which proves the identity (3.5.2), because f is uniquely determined by its orthogonal projections. To prove (ii) and (iii), for fixed θ, we let $B_n(\theta) := c_{\lambda_\kappa} \int_{\theta-1/n}^{\theta+1/n} (\sin t)^{2\lambda_\kappa} dt$ and $g_n(\cos\phi) := 1/B_n(\theta)$ if $|\phi - \theta| \le 1/n$, and $g_n(\cos\phi) := 0$ otherwise. Then

$$(f *_\kappa g_n)(x) = \frac{1}{B_n(\theta)} \int_{\theta-1/n}^{\theta+1/n} T_\phi^\kappa f(x)(\sin\phi)^{2\lambda_\kappa} d\phi.$$

The proof comes down to showing that $f *_\kappa g_n$ converges to f in $L^p(h_\kappa^2, \mathbb{S}^{d-1})$, which can be established first for f being a polynomial and by choosing the polynomial to be, say, the (C,δ)-means of f for $\delta \ge 2\lambda_\kappa + 1$. Finally, (iv) follows from a change of variable in (3.5.2). □

In terms of the generalized translation operator, the maximal function $M_\kappa f$ can be written as

$$\mathscr{M}_\kappa f(x) := \sup_{0<\theta\le\pi} \frac{\int_0^\theta T_\phi^\kappa |f|(x)(\sin\phi)^{2\lambda_\kappa} d\phi}{\int_0^\theta (\sin\phi)^{2\lambda_\kappa} d\phi}. \tag{3.5.4}$$

This relation allows us to prove the following result on the convolution operator.

Theorem 3.5.3. *Assume that $g \in L^1([-1,1], w_{\lambda_\kappa})$ and $|g(\cos\theta)| \le k(\theta)$ for all θ, where $k(\theta)$ is a continuous, nonnegative, and decreasing function on $[0,\pi]$. For $f \in L^1(h_\kappa^2, \mathbb{S}^{d-1})$,*

$$|(f *_\kappa g)(x)| \le c \mathscr{M}_\kappa(|f|)(x), \quad x \in \mathbb{S}^{d-1},$$

where $c = \int_0^\pi k(\theta)(\sin\theta)^{2\lambda_\kappa} d\theta$.

Proof. Let $\lambda = \lambda_\kappa$. Define

$$\Lambda(\theta,x) = \int_0^\theta T_\phi^\kappa |f|(x)(\sin\phi)^{2\lambda}\,d\phi.$$

Then the relation (3.5.4) implies that

$$\Lambda(\theta,x) \le \mathscr{M}_\kappa f(x) \int_0^\theta (\sin\phi)^{2\lambda}\,d\phi$$

for all $x \in \mathbb{S}^{d-1}$. By Proposition 3.5.2,

$$\begin{aligned}|f *_\kappa g(x)| &= c_\lambda \left| \int_0^\pi T_\phi^\kappa f(x) g(\cos\phi)(\sin\phi)^{2\lambda}\,d\phi \right| \\ &\le c_\lambda \int_0^\pi T_\phi^\kappa |f|(x) k(\phi)(\sin\phi)^{2\lambda}\,d\phi.\end{aligned}$$

Integrating by parts, we obtain

$$\begin{aligned}|f *_\kappa g(x)| &\le c_\lambda \left[\Lambda(\pi,x)k(\pi) - \int_0^\pi \Lambda(\theta,x)k'(\theta)d\theta\right] \\ &\le c_\lambda \mathscr{M}_\kappa f(x) \left[k(\pi)\int_0^\pi (\sin\phi)^{2\lambda}\,d\phi - \int_0^\pi k'(\theta)\int_0^\theta (\sin\phi)^{2\lambda}\,d\phi\,d\theta\right],\end{aligned}$$

since $k'(\cos\theta) \le 0$. Integrating by parts again, we conclude that

$$|f *_\kappa g(x)| \le \mathscr{M}_\kappa f(x) c_\lambda \int_0^\pi k(\theta)(\sin\theta)^{2\lambda}\,d\theta \le c\mathscr{M}_\kappa f(x).$$

This completes the proof. □

Applying the above theorem on the Cesàro means gives the following:

Theorem 3.5.4. *If $\delta > \lambda_\kappa$ and $f \in L^1(\mathbb{S}^{d-1}; h_\kappa^2)$, then for every $x \in \mathbb{S}^{d-1}$,*

$$S_*^\delta(x)(f)(x) := \sup_{n\ge 0} |S_n^\delta(h_\kappa^2; f, x)| \le c\left[\mathscr{M}_\kappa f(x) + \mathscr{M}_\kappa f(-x)\right]. \tag{3.5.5}$$

If, in addition, $\delta \ge 2\lambda_k + 1$, then the term $\mathscr{M}_\kappa f(-x)$ in (3.5.5) can be dropped.

Proof. For the proof of the inequality, it suffices to consider the case $\lambda < \delta \le \lambda + 1$, where $\lambda = \lambda_\kappa$, since it is well known that $S_*^{\delta+\tau} f(x) \le S_*^\delta(f)(x)$ for any $\tau > 0$. Setting

$$\begin{aligned}G_{n,1}^\delta(\cos\theta) &:= n^{\lambda-\delta}(n^{-1}+\theta)^{-(\delta+\lambda+1)}\chi_{[0,\frac{\pi}{2}]}(\theta), \\ G_{n,2}^\delta(\cos\theta) &:= n^{\lambda-\delta}(n^{-1}+\theta)^{-\lambda}\chi_{[0,\frac{\pi}{2}]}(\theta),\end{aligned}$$

and using (3.3.8) and the pointwise estimate of the kernel $k_n^\delta(w_\lambda; t, 1)$ for the (C,δ)-means of the Gegenbauer series in [53], we derive that, for $\lambda < \delta \le \lambda + 1$,

$$K_n^\delta(h_\kappa^2; \cos\theta) \le c\left[G_{n,1}^\delta(\cos\theta) + G_{n,2}^\delta(\cos(\pi-\theta))\right].$$

It is easy to see that $g(t) = G^{\delta}_{n,i}(t)$ satisfy the conditions of Theorem 3.5.3, which allows us to conclude that

$$\begin{aligned} |S_n^{\delta} f(x)| &\le \left[(|f| * G^{\delta}_{n,1})(x) + (|f| * G^{\delta}_{n,2})(-x) \right] \\ &\le c\left[\mathscr{M}_{\kappa} f(x) + \mathscr{M}_{\kappa} f(-x)\right]. \end{aligned}$$

Furthermore, if $\delta > 2\lambda + 1$, then the pointwise estimate of the kernel $k_n^{\delta}(w_{\lambda}; t, 1)$ shows that $|K_n^{\delta}(h_{\kappa}^2; \cos\theta)|$ is bounded by a single term and the same proof yields $|S_n^{\delta} f(x)| \le c\mathscr{M} f(x)$. □

3.6 Notes and further results

The first studies on h-harmonic expansions appeared in [69] (which contains (2.3.2) and a proof for the formula for Z_n^{κ}, in the case of $\mathbb{Z}_2^d$, by summing over a specific orthonormal basis using special function identities), and [68] (which contains relations (3.2.3) and (3.2.5), with the latter proved by studying the orthogonal expansion of $V_{\kappa} f$ on the unit ball). The proof of (3.2.5) in Theorem 3.2.3 was given in [16]. The Funk–Hecke formula (3.2.4) was established in [71]. The results on the Cesàro summability for $G = \mathbb{Z}_2^d$ were established in [13, 14, 37], which reduces to the classical results on the spherical harmonics [3, 5, 48] when $\kappa = 0$. The convolution and the translation operators were defined in [72], and used to study weighted best approximation on the sphere. The maximal function $\mathscr{M}_{\kappa} f$ was defined in [73], the results in Section 3.5 were established in [12]. In the case of $G = \mathbb{Z}_2^d$ and $h_{\kappa}(x) = \prod_{i=1}^d |x_i|^{\kappa_i}$, we can consider the weighted Hardy–Littlewood maximal function defined by

$$M_{\kappa} f(x) := \sup_{0<\theta\le\pi} \frac{\int_{c(x,\theta)} |f(y)| h_{\kappa}^2(y) d\sigma(y)}{\int_{c(x,\theta)} h_{\kappa}^2(y) d\sigma(y)}, \tag{3.6.1}$$

for $f \in L^1(\mathbb{S}^{d-1}; h_{\kappa}^2)$ where $c(x,\theta) := \{y \in \mathbb{S}^{d-1} : \arccos\langle x, y\rangle \le \theta\}$, and prove, using the explicit formula for V_{κ}, that $M_{\kappa} f$ dominates the maximal function $\mathscr{M}_{\kappa} f$ defined in Section 3.5,

$$\mathscr{M}_{\kappa} f(x) \le c \sum_{\varepsilon \in \mathbb{Z}_2^d} M_{\kappa} f(x\varepsilon), \qquad x \in \mathbb{S}^{d-1}.$$

The lack of an explicit formula for the intertwining operator V_{κ} has been an obstacle for deriving deeper results relying on the essence of reflection groups. At the moment, little information is known on the intertwining operator for reflection groups other than $\mathbb{Z}_2^d$.

Chapter 4

Littlewood–Paley Theory and the Multiplier Theorem

The main result of this chapter is a Marcinkiewitcz multiplier theorem for h-harmonic expansions. Its proof uses general Littlewood–Paley theory for a symmetric diffusion semi-group. Several Littlewood–Paley type g-functions are introduced and studied via the Cesàro means for h-harmonic expansions. These g-functions provide new equivalent norms for the space $L^p(h_\kappa^2;\mathbb{S}^{d-1})$, and play crucial roles in the proof of the multiplier theorem.

In Section 4.1, several vector-valued inequalities for self-adjoint operators on $L^2(h_\kappa^2;\mathbb{S}^{d-1})$ are established, which will be used in the proof of the main result. A brief description of the general Littlewood–Paley–Stein theory is given in Section 4.2, where a Littlewood–Paley g-function defined via the Poisson semi-group for h-harmonics is introduced and studied as well. The weighted Littlewood–Paley theory on the sphere is developed in Section 4.3, where two new g-functions defined via the Cesàro means play essential roles. With the help of the Littlewood–Paley theory, a Marcinkiewitcz type multiplier theorem for h-harmonic expansions is proved in Section 4.4, which is then applied to obtain a refined Littlewood–Paley inequality in Section 4.5.

4.1 Vector-valued inequalities for self-adjoint operators

Here we establish several vector-valued inequalities, which will play important roles in later sections.

Theorem 4.1.1. *Let $\{T_k\}_{k=0}^\infty$ be a sequence of self-adjoint linear operators on the space $L^2(h_\kappa^2;\mathbb{S}^{d-1})$. Assume that there exists a positive operator $\mathcal{N}$, not necessarily linear, which is bounded on $L^p(h_\kappa^2;\mathbb{S}^{d-1})$ for all $1 < p < \infty$ and pointwisely controls the maximal operator of $\{T_k\}$:*

$$\sup_{k\in\mathbb{Z}_+} |T_k f(x)| \le c\mathcal{N} f(x), \quad \forall x \in \mathbb{S}^{d-1}. \tag{4.1.1}$$

Then for any sequence $\{n_j\}$ of nonnegative integers and any $\{f_j\} \subset L^2(h_\kappa^2;\mathbb{S}^{d-1})$,

$$\Big\|\Big(\sum_{j=0}^{\infty}|T_{n_j}(f_j)|^2\Big)^{1/2}\Big\|_{\kappa,p} \le c'\Big\|\Big(\sum_{j=0}^{\infty}|f_j|^2\Big)^{1/2}\Big\|_{\kappa,p}. \tag{4.1.2}$$

Proof. The proof of (4.1.2) follows the approach of [49, p.104-5], which uses a generalization of the Riesz convexity theorem for sequences of functions. Let $L^p(\ell^q)$ denote the space of all sequences $\{f_k\}$ of functions for which the norm

$$\|(f_k)\|_{L^p(\ell^q)} := \left(\int_{\mathbb{S}^{d-1}}\Big(\sum_{j=0}^{\infty}|f_j(x)|^q\Big)^{p/q} h_\kappa^2(x)d\sigma(x)\right)^{1/p}$$

is finite. If T is a bounded operator on $L^{p_0}(\ell^{q_0})$ and on $L^{p_1}(\ell^{q_1})$, with $1 \le p_0,p_1,q_0,q_1 \le \infty$, then the Riesz convexity theorem states that T is also bounded on $L^{p_t}(\ell^{q_t})$, where

$$\frac{1}{p_t} = \frac{1-t}{p_0}+\frac{t}{p_1}, \quad \frac{1}{q_t} = \frac{1-t}{q_0}+\frac{t}{q_1}, \quad 0 \le t \le 1.$$

We apply this theorem to the operator T mapping the sequence $\{f_j\}$ to the sequence $\{T_{n_j}(f_j)\}$. By (4.1.1), T is bounded on $L^p(\ell^p)$. It is also bounded on $L^p(\ell^\infty)$ since

$$\Big\|\sup_{j\ge 0}|T_{n_j}(f_j)|\Big\|_{\kappa,p} \le \Big\|\mathscr{N}\Big(\sup_{j\ge 0}|f_j|\Big)\Big\|_{\kappa,p} \le c\Big\|\sup_{j\ge 0}|f_j|\Big\|_{\kappa,p}.$$

Hence, the Riesz convexity theorem shows that T is bounded on $L^p(\ell^q)$ if $1 < p \le q \le \infty$. In particular, T is bounded on $L^p(\ell^2)$ if $1 < p \le 2$. The case $2 < p < \infty$ follows by the standard duality argument, since the dual space of $L^p(\ell^2)$ is $L^{p'}(\ell^2)$, where $1/p+1/p' = 1$, under the pairing

$$\langle (f_j),(g_j)\rangle := \int_{\mathbb{S}^{d-1}}\sum_j f_j(x)g_j(x)h_\kappa^2(x)d\sigma(x)$$

and T is self-adjoint under this paring since each T_j is self-adjoint in $L^2(h_\kappa^2;\mathbb{S}^{d-1})$. □

The above proof of Theorem 4.1.1 actually yields the following Fefferman–Stein inequality for the maximal function $\mathscr{M}_\kappa f$ associated to a reflection group.

Corollary 4.1.2. *Let $1 < p \le 2$ and let $\{f_j\}$ be a sequence of functions. Then*

$$\Big\|\Big(\sum_j|\mathscr{M}_\kappa f_j|^2\Big)^{1/2}\Big\|_{\kappa,p} \le c\Big\|\Big(\sum_j|f_j|^2\Big)^{1/2}\Big\|_{\kappa,p}. \tag{4.1.3}$$

As a consequence of Theorem 4.1.1, we also have the following inequality for the Cesàro means $S_n^\delta(h_\kappa)$.

Corollary 4.1.3. *For* $\delta > \lambda_\kappa$, $1 < p < \infty$ *and any sequence* $\{n_j\}$ *of nonnegative integers,*

$$\left\| \left(\sum_{j=0}^{\infty} |S_{n_j}^\delta(h_\kappa^2; f_j)|^2 \right)^{1/2} \right\|_{\kappa,p} \le c \left\| \left(\sum_{j=0}^{\infty} |f_j|^2 \right)^{1/2} \right\|_{\kappa,p}.$$

Proof. The Cesàro means $S_n^\delta(h_\kappa)$ are clearly self-adjoint on $L^2(h_\kappa^2; \mathbb{S}^{d-1})$. By Theorem 4.1.1,

$$\sup_{n \in \mathbb{Z}_+} |S_n^\delta(h_\kappa; f)(x)| \le C\mathscr{M}_\kappa f(x) + C\mathscr{M}_\kappa f(-x), \quad x \in \mathbb{S}^{d-1}, \quad \delta > \lambda_\kappa.$$

Since $\mathscr{M}_\kappa$ is a positive and bounded operator on $L^p(h_\kappa^2, \mathbb{S}^{d-1})$ for all $1 < p < \infty$, the desired conclusion follows directly from Theorem 4.1.1. □

Definition 4.1.4. Let $\eta \in C^\infty(\mathbb{R})$ be such that $\eta(x) = 1$ for $|x| \le 1$ and $\eta(x) = 0$ for $|x| \ge 2$. For $N = 1, 2, \ldots$, we define the operator L_N^κ by

$$L_N^\kappa(f) = \sum_{j=0}^{\infty} \eta\left(\frac{j}{N}\right) \mathrm{proj}_j^\kappa f.$$

The operator L_N^κ is well behaved and has a highly localized kernel in the unweighted case. Similar to the proof of Corollary 4.1.3, we can also deduce the following:

Corollary 4.1.5. *For* $1 < p < \infty$ *and any sequence* $\{n_j\}$ *of nonnegative integers,*

$$\left\| \left(\sum_{j=0}^{\infty} |L_{n_j}^\kappa(f_j)|^2 \right)^{1/2} \right\|_{\kappa,p} \le c \left\| \left(\sum_{j=0}^{\infty} |f_j|^2 \right)^{1/2} \right\|_{\kappa,p}.$$

4.2 The Littlewood–Paley–Stein function

Let (X, μ) be a σ-finite measure space with a positive measure μ. Given $0 < p < \infty$, we denote by $L^p(d\mu)$ the usual Lebesgue space of functions on X with finite quasi-norm

$$\|f\|_{L^p(d\mu)} := \left(\int_X |f(x)|^p \, d\mu(x) \right)^{\frac{1}{p}}.$$

Definition 4.2.1. For a given symmetric diffusion semi-group $\{T_t\}_{t \ge 0}$ on (X, μ), the Littlewood–Paley–Stein function of $f : X \to \mathbb{C}$ is defined by

$$g_0(f)(x) := \left(\int_0^\infty t \left| \frac{\partial}{\partial t} T^t f(x) \right|^2 dt \right)^{\frac{1}{2}}, \quad x \in X. \tag{4.2.1}$$

A general Littlewood–Paley theory for a symmetric diffusion semi-group was developed by E. M. Stein in his 1970 monograph [49]. In particular, the following result was established in [49, Theorem 10, p. 111]:

Theorem 4.2.2. *If $1 < p < \infty$, then for any $f \in L^p(d\mu)$,*

$$\|g_0(f)\|_{L^p(d\mu)} \le c_p \|f\|_{L^p(d\mu)}.$$

If, in addition, $f \in L^1(d\mu) \cap L^p(d\mu)$ and $\int_X f\,d\mu = 0$, then the following inverse inequality is also true:

$$\|f\|_{L^p(d\mu)} \le c_p \|g_0(f)\|_{L^p(d\mu)}.$$

According to Theorem 4.2.2, if $1 < p < \infty$, $f \in L^1(d\mu) \cap L^p(d\mu)$ and $\int_X f\,d\mu = 0$, then

$$\|f\|_{L^p(d\mu)} \sim \|g_0(f)\|_{L^p(d\mu)}.$$

Typical classical examples of the Littlewood–Paley–Stein function are given with the Gauss semi-group $e^{t\Delta}$ and the Poisson semi-group $e^{-t(-\Delta)^{1/2}}$ on the unit circle or $\mathbb{R}^d$, where Δ is the corresponding Laplace operator. The semi-group $e^{-t(-\Delta)^{1/2}}$ is the one originally considered by Littlewood and Paley, and it has the important additional property of being "subordinated" to another symmetric diffusion semi-group, namely $e^{t\Delta}$, in the sense that

$$e^{t\Delta} f = \frac{1}{\sqrt{\pi}} \int_0^\infty \frac{e^{-u}}{\sqrt{u}} e^{t^2\Delta/4u} f\,du.$$

More generally, if $T^t = e^{-tA}$ is a symmetric diffusion semi-group with generator $-A$, and $P_t = e^{-tA^{1/2}}$ (in which case the above integral representation holds with $-A$ in place of Δ), we say that P_t is the subordinated semi-group of T^t.

The above general Littlewood–Paley theory is applicable to the h-harmonic expansions with a symmetric diffusion semi-group defined via Poisson integrals.

Definition 4.2.3. For a function f on the sphere $\mathbb{S}^{d-1}$, define

$$g(f) := \left(\int_0^1 (1-r) \left| \frac{\partial}{\partial r} P_r^\kappa f \right|^2 dr \right)^{\frac{1}{2}}, \tag{4.2.2}$$

where $P_r^\kappa f$ denotes the Poisson integral of f given in (3.4.3).

As an application of Theorem 4.2.2, we have the following result.

Theorem 4.2.4. *If $1 < p < \infty$ and $f \in L^p(h_\kappa^2; \mathbb{S}^{d-1})$, then*

$$\|g(f)\|_{\kappa,p} \le c_{\kappa,p} \|f\|_{\kappa,p}.$$

If, in addition, $\int_{\mathbb{S}^{d-1}} f(x) h_\kappa^2(x)\,d\sigma(x) = 0$, then the following inverse inequality holds:

$$\|f\|_{\kappa,p} \le c_p \|g(f)\|_{\kappa,p}. \tag{4.2.3}$$

Proof. Define $T^t := P_{e^{-t}}^\kappa$ for $t > 0$. Lemma 3.4.5 then implies that T^t, for $t > 0$, is a symmetric diffusion semi-group and hence, using Theorem 4.2.2,

$$c_p^{-1} \|f\|_{\kappa,p} \le \|g_0(f)\|_{\kappa,p} \le c_p \|f\|_{\kappa,p},$$

where the additional condition $\int_{\mathbb{S}^{d-1}} f(x)h_\kappa^2(x)\,d\sigma(x) = 0$ is required in the first inequality. Here, the g_0-function is defined in terms of the Poisson semi-group $P_{e^{-t}}^\kappa$ and, performing the change of variable $r = e^{-t}$, we get

$$g_0(f) := \Big(\int_0^\infty \Big|\frac{\partial}{\partial t}(P_{e^{-t}}^\kappa f)\Big|^2 t\,dt\Big)^{\frac{1}{2}} = \Big(\int_0^1 \Big|\frac{\partial}{\partial r}P_r^\kappa f\Big|^2 r|\log r|\,dr\Big)^{\frac{1}{2}}.$$

To complete the proof, it suffices to show that

$$\|g_0(f)\|_{\kappa,p} \sim \|g(f)\|_{\kappa,p}. \tag{4.2.4}$$

Since $|\log r| \sim 1-r$ as $1/2 \le r \le 1$, and $\lim_{r\to 0^+} r\log r = 0$, we have that

$$g_0(f) \le cg(f) + c\Big(\int_0^{1/2}\Big|\frac{\partial}{\partial r}P_r^\kappa f\Big|^2 dr\Big)^{\frac{1}{2}},$$
$$g(f) \le cg_0(f) + c\Big(\int_0^{1/2}\Big|\frac{\partial}{\partial r}P_r^\kappa f\Big|^2 dr\Big)^{\frac{1}{2}}.$$

However, by Lemma 3.4.5,

$$\sup_{0<r\le 1/2}\Big|\frac{\partial}{\partial r}P_r^\kappa f\Big| \le \sum_{j=1}^\infty j2^{-j+1}|\operatorname{proj}_j^\kappa f|,$$

which, using (3.3.4) and the positivity of V_κ, is bounded above by

$$c\|f\|_{\kappa,1}\sum_{n=1}^\infty n^2 2^{-n+1}\max_{t\in[-1,1]}|C_n^{\lambda_\kappa}(t)| \le c'\|f\|_{\kappa,p} < \infty.$$

Combining these inequalities we get the desired equation (4.2.4). □

4.3 The Littlewood–Paley theory on the sphere

In this section, two new Littlewood–Paley type g-functions are defined and studied via the Cesàro (C, δ)-means for h-harmonic expansions. These g-functions play a central role in the Littlewood–Paley theory on the sphere, and the main motivation for introducing them is to provide new equivalent norms for the spaces $L^p(h_\kappa^2;\mathbb{S}^{d-1})$.

Definition 4.3.1. Given $\delta \ge 0$, the Littlewood–Paley function $g_\delta(f)$ of $f \in L(h_\kappa^2;\mathbb{S}^{d-1})$ is defined by

$$g_\delta(f) = \Big(\sum_{n=1}^\infty |S_n^{\delta+1}(h_\kappa;f) - S_n^\delta(h_\kappa;f)|^2 n^{-1}\Big)^{\frac{1}{2}}. \tag{4.3.1}$$

To define our next Littlewood–Paley function $g_\delta^*(f)$, let $\{v_k\}_{k=1}^\infty$ be an arbitrarily given sequence of positive numbers which satisfies the condition

$$\sup_{n\in\mathbb{N}} n^{-1}\sum_{k=1}^n v_k = M < \infty,$$

where $M > 0$ is a constant. We then fix the sequence $\{v_k\}_{k=1}^\infty$ and define $g_\delta^*(f)$ as follows.

Definition 4.3.2. Given $\delta \ge 0$, the Littlewood–Paley function $g_\delta^*(f)$ of $f \in L^1(h_\kappa^2;\mathbb{S}^{d-1})$ is defined by

$$g_\delta^*(f) := \Big(\sum_{n=1}^{\infty} \big|S_n^{\delta+1}(h_\kappa^2;f) - S_n^\delta(h_\kappa^2;f)\big|^2 n^{-1} v_n\Big)^{\frac{1}{2}}.$$

The main result of this section is the following theorem.

Theorem 4.3.3. *If $f \in L^p(h_\kappa^2;\mathbb{S}^{d-1})$ and $\int_{\mathbb{S}^{d-1}} f(x)h_\kappa^2(x)\,d\sigma(x) = 0$, then*

$$\|f\|_{\kappa,p} \le c_p \|g_\delta(f)\|_{\kappa,p}, \quad \delta \ge 0, \quad 1 < p < \infty.$$

Conversely, if the vector-valued inequality,

$$\Big\|\Big(\sum_{j=1}^{\infty} |S_{n_j}^\delta(h_\kappa;f_j)|^2\Big)^{\frac{1}{2}}\Big\|_{\kappa,p} \le c_p \Big\|\Big(\sum_{j=1}^{\infty} |f_j|^2\Big)^{\frac{1}{2}}\Big\|_{\kappa,p}, \quad \forall \{n_j\} \subset \mathbb{N}, \tag{4.3.2}$$

holds for all sequences $\{f_j\} \subset L(h_\kappa^2;\mathbb{S}^{d-1})$, then the following inverse inequality holds:

$$\|g_\delta^*(f)\|_{\kappa,p} \le c_p M \|f\|_{\kappa,p}, \quad \forall f \in L^p(h_\kappa^2;\mathbb{S}^{d-1}),$$

where the constant c_p is independent of the sequence $\{v_n\}_{n=1}^\infty$.

Note that if, in particular, we choose $v_j = 1$ for all $j \in \mathbb{N}$, then $g_\delta^*(f) = g_\delta(f)$. Thus, for the Littlewood–Paley function $g_\delta(f)$, we have the following immediate corollary.

Corollary 4.3.4. *If $1 < p < \infty$, $f \in L^p(h_\kappa^2;\mathbb{S}^{d-1})$ and $\delta > \lambda_\kappa$, then*

$$c_p^{-1}\|f\|_{\kappa,p} \le \|g_\delta(f)\|_{\kappa,p} \le c_p \|f\|_{\kappa,p},$$

where the additional condition $\int_{\mathbb{S}^{d-1}} f(x)h_\kappa^2(x)\,d\sigma(x) = 0$ is required in the first inequality.

The proof of Theorem 4.3.3 is given in the following two subsections.

4.3.1 A crucial lemma

The proof of Theorem 4.3.3 requires a crucial lemma, which is proven in this subsection. We denote by $|I|$ the length of a given interval $I \subset \mathbb{R}$.

Lemma 4.3.5. *Assume that $1 < p < \infty$, $\delta \ge 0$ and*

$$\Big\|\Big(\sum_{j=1}^{\infty} |S_{n_j}^\delta(h_\kappa;f_j)|^2\Big)^{\frac{1}{2}}\Big\|_{\kappa,p} \le c_p \Big\|\Big(\sum_{j=1}^{\infty} |f_j|^2\Big)^{\frac{1}{2}}\Big\|_{\kappa,p}, \quad \forall \{n_j\} \subset \mathbb{N} \tag{4.3.3}$$

for any sequence of functions $\{f_j\} \subset L(h_\kappa^2;\mathbb{S}^{d-1})$. If $r_j \in (0,1)$ and I_j is a subinterval of $[r_j,1)$ for $j = 1,2,\dots$, then

$$\Big\|\Big(\sum_{j=1}^{\infty} |S_{n_j}^\delta(h_\kappa;P_{r_j}^\kappa f_j)|^2\Big)^{\frac{1}{2}}\Big\|_{\kappa,p} \le c_p \Big\|\Big(\sum_{j=1}^{\infty} \frac{1}{|I_j|}\int_{I_j} |P_r^\kappa f_j|^2\,dr\Big)^{\frac{1}{2}}\Big\|_{\kappa,p}, \tag{4.3.4}$$

for all $\{n_j\}_{k=1}^\infty \subset \mathbb{N}$ and $\{f_j\}_{k=1}^\infty \subset L(h_\kappa^2;\mathbb{S}^{d-1})$ with the constant c_p being independent of $\{r_j\}$, $\{I_j\}$, $\{n_j\}$ and $\{f_j\}$.

Proof. For simplicity, we shall write $S_n^\delta f$ for the Cesàro (C,δ)-means $S_n^\delta(h_\kappa; f)$ in the proof below. We first claim that for each $\{n_j\} \subset \mathbb{N}$ and $\{f_j\} \subset L(h_\kappa^2; \mathbb{S}^{d-1})$,

$$\Big\| \Big(\sum_{j=1}^{\infty} |S_{n_j}^\delta P_{r_j}^\kappa f_j|^2 \Big)^{\frac{1}{2}} \Big\|_{\kappa,p} \le c_p \Big\| \Big(\sum_{j=1}^{\infty} |f_j|^2 \Big)^{\frac{1}{2}} \Big\|_{\kappa,p}. \tag{4.3.5}$$

To see this, we use Lemma 4.3.7 below to obtain

$$|S_{n_j}^\delta P_{r_j}^\kappa f_j|^2 \le c \sum_{\ell=0}^{n_j} |b_{\ell,n_j}^\delta| |S_\ell^\delta f_j|^2, \quad j = 1, 2, \dots.$$

Summing over j and invoking (4.3.3), we deduce

$$\begin{aligned} \Big\| \Big(\sum_{j=1}^{\infty} |S_{n_j}^\delta P_{r_j}^\kappa f_j|^2 \Big)^{\frac{1}{2}} \Big\|_{\kappa,p} &\le c_p \Big\| \Big(\sum_{j=1}^{\infty} \sum_{\ell=0}^{n_j} |b_{\ell,n_j}^\delta| |f_j|^2 \Big)^{\frac{1}{2}} \Big\|_{\kappa,p} \\ &\le c \Big\| \Big(\sum_{j=1}^{\infty} |f_j|^2 \Big)^{\frac{1}{2}} \Big\|_{\kappa,p}, \end{aligned}$$

which proves (4.3.5).

Next, we show that the desired inequality (4.3.4) follows from (4.3.5). To see this, for each $j \ge 0$ and $n \ge 1$, we let $\{r_{j,i}\}_{i=0}^{2^n} \subset I_j$ be such that $r_{j,i} - r_{j,i-1} = 2^{-n}|I_j|$ for all $1 \le i \le 2^n$. Then, for each $n \in \mathbb{N}$, $R_{j,n} := 2^{-n} \sum_{i=1}^{2^n} |P_{r_{j,i}}^\kappa f_j|^2$ is a Riemann sum of the integral $\frac{1}{|I_j|} \int_{I_j} |P_r^\kappa f_j|^2 \, dr$. Thus, by Fatou's theorem, it follows that

$$\Big\| \Big(\sum_{j=1}^{\infty} \frac{1}{|I_j|} \int_{I_j} |P_r^\kappa f_j|^2 \, dr \Big)^{\frac{1}{2}} \Big\|_{\kappa,p} = \lim_{n\to\infty} \Big\| \Big(2^{-n} \sum_{j=1}^{\infty} \sum_{i=1}^{2^n} |P_{r_{j,i}}^\kappa f_j|^2 \Big)^{\frac{1}{2}} \Big\|_{\kappa,p}.$$

On the other hand, since for each fixed $n \in \mathbb{N}$, $r_j < r_{j,i}$ for all $1 \le i \le n$ and $j \in \mathbb{N}$, using (4.3.5) we have

$$\begin{aligned} \Big\| \Big(\sum_{j=1}^{\infty} |S_{n_j}^\delta P_{r_j}^\kappa f_j|^2 \Big)^{\frac{1}{2}} \Big\|_{\kappa,p} &= \Big\| \Big(2^{-n} \sum_{i=1}^{2^n} \sum_{j=1}^{\infty} |S_{n_j}^\delta P_{r_j/r_{j,i}}^\kappa (P_{r_{j,i}}^\kappa f_j)|^2 \Big)^{\frac{1}{2}} \Big\|_{\kappa,p} \\ &\le c_p \Big\| \Big(2^{-n} \sum_{i=1}^{2^n} \sum_{j=1}^{\infty} |S_{n_j}^\delta (P_{r_{j,i}}^\kappa f_j)|^2 \Big)^{\frac{1}{2}} \Big\|_{\kappa,p}. \end{aligned}$$

Thus, letting $n \to \infty$, we obtain (4.3.4), and this completes the proof of the lemma. □

The proof of the main theorem needs two more elementary lemmas. We introduce the following notation: let s_j^δ, $j = 0, 1, \dots$, denote the Cesàro (C, δ)-means of a given sequence $\{a_j\}_{j=0}^\infty$ of complex numbers. For a given $r \in (0,1)$, define

$$s_{n,r}^\delta := \sum_{j=0}^{n} \frac{A_{n-j}^\delta}{A_n^\delta} a_j r^j.$$

Lemma 4.3.6. *If $\delta \ge 0$, $n \in \mathbb{N}$, and $1 - n^{-1} \le r < 1$, then*

$$s_n^\delta = r^{-n} s_{n,r}^\delta + \sum_{j=0}^{n-1} a_{j,n}^\delta s_{j,r}^\delta,$$

where $a_{j,n}^\delta$ are constants independent of the sequence $\{a_j\}$ and satisfying

$$\max_{0 \le j \le n-1} |a_{j,n}^\delta| \le c_\delta (1-r).$$

Lemma 4.3.7. *If $\delta \ge 0$ and $0 < r < 1$, then*

$$s_{n,r}^\delta = \sum_{j=0}^{n} b_{j,n}^\delta s_j^\delta, \quad n = 1, 2, \ldots,$$

where $b_{j,n}^\delta$ are constants independent of $\{a_j\}$ and satisfying

$$\sum_{j=0}^{n} |b_{j,n}^\delta| \le c_\delta.$$

The proofs of these two lemmas can be found in [3] and [16, Section 3.2], respectively.

4.3.2 Proof of Theorem 4.3.3

For simplicity, we shall write $S_n^\delta f$ for the Cesàro (C,δ)-means $S_n^\delta(h_\kappa; f)$ in the proof below. We first prove the inequality $\|f\|_{\kappa,p} \le c_p \|g_\delta(f)\|_{\kappa,p}$. By Theorem 4.2.4, it suffices to show that

$$g(f)(x) \le c g_\delta(f)(x), \quad \forall x \in \mathbb{S}^{d-1}. \tag{4.3.6}$$

To see this, we note first that

$$\begin{aligned} \left|\frac{\partial}{\partial r} P_r^\kappa f\right| &= (1-r)^{\delta+1}(1-r)^{-\delta-1}\left|\sum_{j=0}^{\infty} j r^{j-1} \operatorname{proj}_j^\kappa f\right| \\ &= (1-r)^{\delta+1}\left|\sum_{n=1}^{\infty}\Big(\sum_{j=0}^{n} j A_{n-j}^\delta \operatorname{proj}_j^\kappa f\Big) r^{n-1}\right|. \end{aligned}$$

Since a quick computation shows that

$$S_n^{\delta+1} f - S_n^\delta f = -(n+\delta+1)^{-1}(A_n^\delta)^{-1} \sum_{j=0}^{n} j A_{n-j}^\delta \operatorname{proj}_j^\kappa f,$$

it follows that

$$\left|\frac{\partial}{\partial r} P_r^\kappa f\right| \le c(1-r)^{\delta+1} \sum_{n=1}^{\infty} n A_n^\delta |S_n^{\delta+1} f - S_n^\delta f| r^{n-1},$$

which, by the Cauchy–Schwartz inequality, implies

$$\begin{aligned}\left|\frac{\partial}{\partial r}P_r^\kappa f\right|^2 &\le c(1-r)^{2\delta+2}\Big(\sum_{n=1}^\infty nA_n^\delta\left|S_n^{\delta+1}f-S_n^\delta f\right|^2 r^{n-1}\Big)\Big(\sum_{n=1}^\infty nA_n^\delta r^{n-1}\Big)\\ &= c(1+\delta)(1-r)^\delta\sum_{n=1}^\infty nA_n^\delta|S_n^{\delta+1}f-S_n^\delta f|^2 r^{n-1}.\end{aligned}$$

Consequently,

$$\begin{aligned}|g(f)|^2 &= \int_0^1\left|\frac{\partial}{\partial r}P_r^\kappa f\right|^2(1-r)\,dr \le c\sum_{n=1}^\infty nA_n^\delta|S_n^{\delta+1}f-S_n^\delta f|^2\int_0^1(1-r)^{1+\delta}r^{n-1}\,dr\\ &\le c\sum_{n=1}^\infty n^{-1}|S_n^{\delta+1}f-S_n^\delta f|^2 = |g_\delta(f)|^2,\end{aligned}$$

where the third step uses the fact that $\int_0^1(1-r)^{\delta+1}r^{n-1}\,dr = \frac{\Gamma(\delta+2)\Gamma(n)}{\Gamma(n+\delta+2)} \sim n^{-\delta-2}$. This proves the desired inequality (4.3.6).

We now turn to the proof of the second assertion in Theorem 4.3.3. Without loss of generality, we may assume that $n \le \sum_{j=1}^n v_j \le 2n$, since the desired conclusion for general $\{v_j\}$ can be deduced by applying the result in this special case to the two sequences $\widetilde{v}_j = 1$ and $\widetilde{v}_j = M^{-1}v_j + 1$, respectively. For convenience, we define, for $n = 1, 2, \dots$,

$$E_n f = -(n+1+\delta)^{-1}\sum_{j=0}^n j\operatorname{proj}_j^\kappa f.$$

It is easily seen that, for $0 \le j \le n$,

$$S_j^\delta(E_n f) = \frac{j+\delta+1}{n+\delta+1}\Big(S_j^{\delta+1}f - S_j^\delta f\Big). \tag{4.3.7}$$

Using Lemma 4.3.6 we obtain that, for any $r \in [1-n^{-1}, 1)$,

$$\begin{aligned}S_n^{\delta+1}f - S_n^\delta f &= S_n^\delta(E_n f) = r^{-n}P_r^\kappa(S_n^\delta(E_n f)) + \sum_{j=1}^{n-1}a_{j,n}^\delta P_r^\kappa(S_j^\delta(E_n f))\\ &= r^{-n}\Big(S_n^{\delta+1}(P_r^\kappa f) - S_n^\delta(P_r^\kappa f)\Big) + \sum_{j=1}^{n-1}\frac{j+\delta+1}{n+\delta+1}a_{j,n}^\delta\Big[S_j^{\delta+1}(P_r^\kappa f) - S_j^\delta(P_r^\kappa f)\Big],\end{aligned} \tag{4.3.8}$$

where $|a_{j,n}^\delta| \le c(1-r) \le cn^{-1}$, and the last step uses (4.3.7). Now let $\mu_1 = 1$, and $\mu_n = 1 + \sum_{i=1}^{n-1} v_i$ for $n > 1$. Clearly, $r_n := 1 - \frac{1}{\mu_n} \in [1-n^{-1}, 1-(2n-1)^{-1}]$. Thus, applying (4.3.8) with $r = r_n$, and setting $f_n = P_{r_n}^\kappa f$, we deduce that

$$|S_n^{\delta+1}f - S_n^\delta f| \le c|S_n^{\delta+1}f_n - S_n^\delta f_n| + cn^{-2}\sum_{j=1}^{n-1} j|S_j^{\delta+1}(f_n) - S_j^\delta(f_n)|,$$

which, using the Cauchy–Schwartz inequality again, yields

$$|S_n^{\delta+1} f - S_n^\delta f|^2 \le c|S_n^{\delta+1} f_n - S_n^\delta f_n|^2 + cn^{-3} \sum_{j=1}^{n-1} j^2 |S_j^{\delta+1} f_n - S_j^\delta f_n|^2. \tag{4.3.9}$$

Therefore, by (4.3.9) and Corollary 4.2.4, to complete the proof of the second assertion in Theorem 4.3.3 it suffices to show the inequalities

$$\Big\| \Big(\sum_{n=1}^\infty n^{-1} |S_n^{\delta+1} f_n - S_n^\delta f_n|^2 \nu_n \Big)^{\frac{1}{2}} \Big\|_{\kappa,p} \le c_p \|g(f)\|_{\kappa,p} \tag{4.3.10}$$

and

$$\Big\| \Big(\sum_{n=1}^\infty \frac{\nu_n}{n^4} \sum_{j=1}^{n-1} j^2 |S_j^{\delta+1} f_n - S_j^\delta f_n|^2 \Big)^{\frac{1}{2}} \Big\|_{\kappa,p} \le c_p \|g(f)\|_{\kappa,p}. \tag{4.3.11}$$

To this end, let $\eta \in C^\infty(\mathbb{R})$ and L_n^κ be as in Definition 4.1.4. Define

$$D_n f = -\sum_{j=0}^{2n} j\eta\Big(\frac{j}{n}\Big) \operatorname{proj}_j^\kappa f, \quad n = 1, 2, \ldots,$$

and observe that, for $1 \le j \le n \le N$,

$$S_j^{\delta+1} f_n - S_j^\delta f_n = (j+\delta+1)^{-1} P_{r_n}^\kappa (S_j^\delta (D_N f)). \tag{4.3.12}$$

Thus, using Lemma 4.3.5 and (4.3.12) with $j = n$, we obtain

$$\begin{aligned}
\Big\| \Big(\sum_{n=1}^N n^{-1} |S_n^{\delta+1} f_n - S_n^\delta f_n|^2 \nu_n \Big)^{\frac{1}{2}} \Big\|_{\kappa,p} &\le c \Big\| \Big(\sum_{n=1}^N \frac{\nu_n}{n^3} |P_{r_n}^\kappa (S_n^\delta (D_N f))|^2 \Big)^{\frac{1}{2}} \Big\|_{\kappa,p} \\
&\le c \Big\| \Big(\sum_{n=1}^N \frac{\nu_n}{n^3} \frac{1}{r_{n+1} - r_n} \int_{r_n}^{r_{n+1}} |P_r^\kappa (D_N f)|^2 \, dr \Big)^{\frac{1}{2}} \Big\|_{\kappa,p}.
\end{aligned}$$

Since

$$|P_r^\kappa (D_N f)| = r \Big| L_N^\kappa \Big(\frac{\partial}{\partial r} P_r^\kappa f \Big) \Big|,$$

applying Corollary 4.1.5 to the Riemann sums of the integrals $\int_{r_n}^{r_{n+1}}$, we obtain

$$\begin{aligned}
&\Big\| \Big(\sum_{n=1}^N \frac{\nu_n}{n^3} \frac{1}{r_{n+1} - r_n} \int_{r_n}^{r_{n+1}} |P_r^\kappa (D_N f)|^2 \, dr \Big)^{\frac{1}{2}} \Big\|_{\kappa,p} \\
&\qquad \le c_p \Big\| \Big(\sum_{n=1}^N \frac{\nu_n}{n^3} \frac{1}{r_{n+1} - r_n} \int_{r_n}^{r_{n+1}} \Big| \frac{\partial}{\partial r} P_r^\kappa f \Big|^2 dr \Big)^{\frac{1}{2}} \Big\|_{\kappa,p}.
\end{aligned}$$

Since $r_{n+1} - r_n = \frac{\nu_n}{\mu_n \mu_{n+1}} \sim \frac{\nu_n}{n^2}$ and $1 - r \sim \frac{1}{n}$ for all $r \in [r_n, r_{n+1}]$, it follows that

$$\begin{aligned}
\Big\| \Big(\sum_{n=1}^N \frac{\nu_n}{n^3} \frac{1}{r_{n+1} - r_n} \int_{r_n}^{r_{n+1}} \Big| \frac{\partial}{\partial r} P_r^\kappa f \Big|^2 dr \Big)^{\frac{1}{2}} \Big\|_{\kappa,p} &\le c_p \Big\| \Big(\sum_{n=1}^\infty \frac{1}{n} \int_{r_n}^{r_{n+1}} \Big| \frac{\partial}{\partial r} P_r^\kappa f \Big|^2 dr \Big)^{\frac{1}{2}} \Big\|_{\kappa,p} \\
&\le c_p \|g(f)\|_{\kappa,p}.
\end{aligned}$$

Putting the above together, and letting $N \to \infty$, we obtain (4.3.10).

The proof of (4.3.11) is similar. In fact, using Lemma 4.3.5 and (4.3.12), we have

$$\Big\|\Big(\sum_{n=1}^{N}\frac{v_n}{n^4}\sum_{j=1}^{n-1} j^2|S_\ell^{\delta+1}f_n - S_\ell^\delta f_n|^2\Big)^{\frac12}\Big\|_{\kappa,p} \le c\Big\|\Big(\sum_{n=1}^{N}\frac{v_n}{n^4}\sum_{j=1}^{n-1}|P_{r_n}^\kappa(S_j^\delta D_N f)|^2\Big)^{\frac12}\Big\|_{\kappa,p}$$

$$\le c_p\Big\|\Big(\sum_{n=1}^{\infty}\frac{v_n}{n^4}\sum_{j=1}^{n-1}\frac{1}{r_{n+1}-r_n}\int_{r_n}^{r_{n+1}}|\frac{\partial}{\partial r}P_r^\kappa f|^2\,dr\Big)^{\frac12}\Big\|_{\kappa,p}$$

$$\le c_p\Big\|\Big(\sum_{n=1}^{\infty}\frac{v_n}{n^3}\frac{1}{r_{n+1}-r_n}\int_{r_n}^{r_{n+1}}|\frac{\partial}{\partial r}P_r^\kappa f|^2\,dr\Big)^{\frac12}\Big\|_{\kappa,p} \le c_p\|g(f)\|_{\kappa,p}.$$

Letting $N \to \infty$ yields (4.3.11). □

4.4 The Marcinkiewicz type multiplier theorem

In this section, we prove the Marcinkiewicz type multiplier theorem for spherical h-harmonic expansions. The conditions on the multiplier are stated in terms of the forward difference, defined below.

Definition 4.4.1. Given a sequence $\{a_j\}_{j=0}^\infty$ of complex numbers, define

$$\triangle a_j = a_j - a_{j+1}, \quad \triangle^{n+1}a_j = \triangle^n a_j - \triangle^n a_{j+1}.$$

Theorem 4.4.2. *Let $\{\mu_j\}_{j=0}^\infty$ be a sequence of complex numbers satisfying*

(A_0) $\quad \sup_j |\mu_j| \le M < \infty,$

(A_{n_0}) $\quad \sup_{j\ge1} 2^{j(n_0-1)}\sum_{l=2^j}^{2^{j+1}}|\triangle^{n_0}\mu_l| \le M < \infty,$

where n_0 is the smallest integer bigger than or equal to $\lambda_\kappa + 1$. Then $\{\mu_j\}$ defines an $L^p(h_\kappa^2;\mathbb{S}^{d-1})$ multiplier for all $1 < p < \infty$; namely,

$$\Big\|\sum_{j=0}^{\infty}\mu_j \operatorname{proj}_j^\kappa f\Big\|_{\kappa,p} \le c_p M\|f\|_{\kappa,p}, \qquad 1<p<\infty,$$

where the constant c_p is independent from $\{\mu_j\}$ and f.

Remark 4.4.3. If $\{\mu_j\}$ is a sequence of complex numbers satisfying the condition (A_k) for some positive integer k, then, as can be easily verified from the definition, $\{\mu_j\}$ satisfies the condition (A_i) for all $1 \le i \le k$, with a possibly different absolute constant M.

The proof of Theorem 4.4.2 relies on the following lemma, whose proof can be found in [3] and [16, Lemma 3.3.3].

Lemma 4.4.4. *Let $\{\mu_j\}_{j=0}^{\infty}$ be a bounded sequence of complex numbers satisfying the condition $(A_{\delta+1})$ for some nonnegative integer δ. Assume that $\{a_j\}_{j=0}^{\infty}$ is another sequence of complex numbers. Let s_n^δ and σ_n^δ denote the Cesàro (C,δ)-means of the sequences $\{a_j\}_{j=0}^{\infty}$ and $\{a_j\mu_j\}_{j=0}^{\infty}$, respectively. Then*

$$\sigma_n^\delta = \mu_n s_n^\delta + \sum_{\ell=0}^{n-1} C_{\ell,n}^\delta s_\ell^\delta,$$

where the constants $C_{\ell,N}^\delta$ are independent of $\{a_j\}_{j=0}^{\infty}$ and satisfy

$$\left|C_{\ell,n}^\delta\right| \le c\sum_{k=1}^{\delta+1}(\ell+1)^{k-1}\left|\triangle^k\mu_\ell\right|, \qquad \ell=0,1,\dots,n-1. \tag{4.4.1}$$

Proof of Theorem 4.4.2. Without loss of generality, we may assume $\mu_0=0$ and $M=1$. Let δ be the smallest integer bigger than λ_κ. Then (4.3.2) follows from Theorem 4.1.3. Now let $F=\sum_{j=1}^{\infty}\mu_j \operatorname{proj}_j^\kappa f$. By Theorem 4.3.3 and Corollary 4.3.4, it suffices to show

$$g_\delta(F) \le c\Big(\sum_{n=1}^{\infty}|S_n^{\delta+1}f - S_n^\delta f|^2 v_n n^{-1}\Big)^{\frac12} =: g_\delta^*(f), \tag{4.4.2}$$

for every sequence of positive numbers $\{v_n\}$ satisfying the condition

$$\sup_{N\in\mathbb{N}} N^{-1}\sum_{j=1}^{N} v_j \le cM < \infty.$$

For the proof of (4.4.2), we use Lemma 4.4.4 to obtain

$$\begin{aligned} S_n^{\delta+1}F - S_n^\delta F &= \frac{-1}{n+\delta+1}(A_n^\delta)^{-1}\sum_{j=0}^{n} A_{n-j}^\delta \mu_j j \operatorname{proj}_j^\kappa f \\ &= \frac{1}{n+\delta+1}\Big(\mu_n\sigma_n^\delta + \sum_{\ell=0}^{n-1} C_{\ell,n}^\delta \sigma_\ell^\delta\Big), \end{aligned}$$

where

$$\sigma_\ell^\delta = -(A_\ell^\delta)^{-1}\sum_{j=0}^{\ell} A_{\ell-j}^\delta j \operatorname{proj}_j^\kappa f = (\ell+\delta+1)\big(S_\ell^{\delta+1}f - S_\ell^\delta f\big).$$

It then follows by (4.4.1) that

$$|S_n^{\delta+1}F - S_n^\delta F| \le |\mu_n||S_n^{\delta+1}f - S_n^\delta f| + Cn^{-1}\sum_{j=1}^{\delta+1}\sum_{\ell=1}^{n-1}\ell^j|\triangle^j\mu_\ell||S_\ell^{\delta+1}f - S_\ell^\delta f|.$$

On the other hand, using Remark 4.4.3, we deduce from condition $(A_{\delta+1})$ that

$$\sum_{j=1}^{\delta+1}\sum_{\ell=1}^{n-1}\ell^j|\triangle^j\mu_\ell| \le cn. \tag{4.4.3}$$

Thus, using (4.4.3) and the Cauchy–Schwartz inequality,

$$\begin{aligned}|g_\delta(F)|^2 &\le c|g_\delta(f)|^2 + c\sum_{j=1}^{\delta+1}\sum_{n=1}^{\infty} n^{-2}\Big(\sum_{\ell=1}^{n-1}\ell^j|\triangle^j\mu_\ell||S_\ell^{\delta+1}f - S_\ell^\delta f|^2\Big)\\ &\le c|g_\delta(f)|^2 + c\sum_{\ell=1}^{\infty}|S_\ell^{\delta+1}f - S_\ell^\delta f|^2\sum_{j=1}^{\delta+1}\ell^{j-1}|\triangle^j\mu_\ell|\\ &\le c\sum_{n=1}^{\infty}|S_n^{\delta+1}f - S_n^\delta f|^2 v_n n^{-1},\end{aligned}$$

where $v_n = 1+\sum_{j=1}^{\delta+1}|\triangle^j\mu_n|n^j$. However, using (4.4.3) once again,

$$n^{-1}\sum_{\ell=1}^{n} v_\ell = 1+\sum_{j=1}^{\delta+1} n^{-1}\sum_{\ell=1}^{n}\ell^j|\triangle^j\mu_\ell| \le c.$$

This proves the desired equation (4.4.2) and completes the proof of Theorem 4.4.2. □

4.5 A Littlewood–Paley inequality

In this section, as an application of the Marcinkiewicz multiplier theorem, we prove a useful Littlewood–Paley type inequality. Let us start with the following definition.

Definition 4.5.1. Given a compactly supported continuous function $\theta : [0,\infty) \to \mathbb{R}$, we define a sequence of operators $\Delta_{\theta,j}^\kappa$ by $\Delta_{\theta,0}^\kappa(f) = \operatorname{proj}_0^\kappa(f)$, and

$$\Delta_{\theta,j}^\kappa(f) := \sum_{n=0}^{\infty}\theta\left(\frac{n}{2^j}\right)\operatorname{proj}_n^\kappa(f), \quad j = 1,2,\ldots.$$

Theorem 4.5.2. *Let m be the smallest positive integer greater than $\lambda_\kappa + 1$. If θ is a compactly supported function in $C^m[0,\infty)$ with* $\operatorname{supp}\theta \subset (a,b)$ *for some $0 < a < b < \infty$, then, for all $f \in L^p(h_\kappa^2;\mathbb{S}^{d-1})$ with $1 < p < \infty$,*

$$\Big\|\Big(\sum_{j=0}^{\infty}|\Delta_{\theta,j}^\kappa f|^2\Big)^{1/2}\Big\|_{\kappa,p} \le c\|f\|_{\kappa,p}, \tag{4.5.1}$$

where c depends only on p,d,κ,a and b. If, in addition,

$$0 < A_1 \le \sum_{j=0}^{\infty}|\theta(2^{-j}t)|^2 \le A_2 < \infty, \quad \forall t > 0, \tag{4.5.2}$$

and $\int_{\mathbb{S}^{d-1}} f(x)h_\kappa^2(x)\,d\sigma(x) = 0$, then

$$\Big\|\Big(\sum_{j=0}^{\infty}|\Delta_{\theta,j}^\kappa f|^2\Big)^{1/2}\Big\|_{\kappa,p} \sim \|f\|_{\kappa,p}, \quad 1 < p < \infty. \tag{4.5.3}$$

Proof. Firstly, let us prove (4.5.1). Let $\{\xi_j\}_{j=0}^{\infty}$ be a sequence of independent random variables taking the values ± 1 and having zero mean. Then, by the Khinchine inequality, for any sequence $\{a_j\}$ of complex numbers,

$$\left(\mathbb{E}\left|\sum_{j=0}^{\infty} a_j \xi_j\right|^p\right)^{1/p} \sim \left(\sum_{j=0}^{\infty} |a_j|^2\right)^{\frac{1}{2}}, \quad 0 < p < \infty, \tag{4.5.4}$$

where $\mathbb{E}$ denotes the expectation of random variables. Now consider the (random) linear operator

$$Tf = \sum_{j=0}^{\infty} \xi_j \Delta_{\theta,j}^{\kappa} f. \tag{4.5.5}$$

Directly from the definition of $\Delta_{\theta,j}^{\kappa} f$, Tf can be rewritten in the form

$$Tf = \sum_{k=1}^{\infty} A(k) \operatorname{proj}_k^{\kappa} f, \quad A(u) := \sum_{j=0}^{\infty} \theta\left(\frac{u}{2^j}\right) \xi_j.$$

Since $\theta \in C^m[0,\infty)$ is supported in a finite interval $(a,b) \subset (0,\infty)$, it follows by a straightforward computation that

$$\left|\left(\frac{d}{du}\right)^r A(u)\right| \le c_r u^{-r}, \quad u \ge 1, \quad r = 0,1,\ldots,m,$$

which, in particular, implies that

$$|\triangle^r A(k)| \le c_r' k^{-r}, \quad r = 0,1,\ldots m, \quad k \ge 1,$$

where the constants c_r and c_r' are independent of the random variables ξ_j. We now apply the Marcinkiewicz multiplier theorem (Theorem 4.4.2) with $\mu_k = A(k)$ to deduce that

$$\|Tf\|_{\kappa,p} \le c_p \|f\|_{\kappa,p}, \quad 1 < p < \infty, \tag{4.5.6}$$

where c_p is a constant depending only on p, d and κ. Combining (4.5.4) and (4.5.5) with (4.5.6), we conclude that

$$\left\|\left(\sum_{j=0}^{\infty} |\Delta_{\theta,j}^{\kappa} f|^2\right)^{\frac{1}{2}}\right\|_{\kappa,p} \sim \left(\mathbb{E}\|Tf\|_{\kappa,p}^p\right)^{1/p} \le c_p \|f\|_{\kappa,p},$$

which proves the desired inequality (4.5.1).

Secondly, we prove the inverse inequality

$$\left\|\left(\sum_{j=1}^{\infty} (\Delta_{\theta,j}^{\kappa} f)^2\right)^{1/2}\right\|_{\kappa,p} \ge c_p' \|f\|_{\kappa,p}, \tag{4.5.7}$$

for $f \in L^p(h_\kappa^2;\mathbb{S}^{d-1})$ with $1 < p < \infty$ and $\int_{\mathbb{S}^{d-1}} f(x)\, h_\kappa^2(x)\, d\sigma(x) = 0$ under the additional assumption that

$$\sum_{j=0}^{\infty} \left|\theta(2^{-j}x)\right|^2 = 1, \quad x > 0. \tag{4.5.8}$$

This assumption implies that, for every spherical polynomial g,

$$\sum_{j=0}^{\infty} (\Delta_{\theta,j}^\kappa \circ \Delta_{\theta,j}^\kappa) g = g - \operatorname{proj}_0^\kappa g. \tag{4.5.9}$$

Now, for $f \in L^p(h_\kappa^2;\mathbb{S}^{d-1})$ with $\int_{\mathbb{S}^{d-1}} f(x) h_\kappa^2(x)\, d\sigma(x) = 0$ and $\varepsilon > 0$, there is a $g \in L^q(h_\kappa^2;\mathbb{S}^{d-1})$ with $\|g\|_{\kappa,1} = 1$, where $\frac{1}{p} + \frac{1}{q} = 1$, such that

$$\|f\|_{\kappa,p} - \varepsilon/2 \le \frac{1}{\omega_d} \int_{\mathbb{S}^{d-1}} f(x) g(x) h_\kappa^2(x)\, d\sigma(x).$$

Let g_n be a spherical polynomial such that $\|g - g_n\|_{\kappa,p} < \varepsilon/2$. Then it follows readily that $\|f\|_{\kappa,p} - \varepsilon \le \frac{1}{\omega_d} \left| \int_{\mathbb{S}^{d-1}} f g_n h_\kappa^2\, d\sigma(x) \right|$. Using (4.5.9), we have

$$\begin{aligned}
\frac{1}{\omega_d}\left|\int_{\mathbb{S}^{d-1}} f g_n h_\kappa^2\, d\sigma(x)\right| &= \frac{1}{\omega_d}\left|\int_{\mathbb{S}^{d-1}} \sum_{j=0}^{\infty} \Delta_{\theta,j}^\kappa f(x) \Delta_{\theta,j}^\kappa g_n(x) h_\kappa^2(x)\, d\sigma(x)\right| \\
&\le c \left\| \left(\sum_{j=0}^{\infty} |\Delta_{\theta,j}^\kappa f|^2 \right)^{\frac{1}{2}} \right\|_{\kappa,p} \left\| \left(\sum_{j=0}^{\infty} |\Delta_{\theta,j}^\kappa g_n|^2 \right)^{\frac{1}{2}} \right\|_{\kappa,q} \\
&\le c \|g_n\|_{\kappa,q} \left\| \left(\sum_{j=0}^{\infty} |\Delta_{\theta,j}^\kappa f(x)|^2 \right)^{\frac{1}{2}} \right\|_{\kappa,p} \le c \left\| \left(\sum_{j=0}^{\infty} |\Delta_{\theta,j}^\kappa f(x)|^2 \right)^{\frac{1}{2}} \right\|_{\kappa,p}.
\end{aligned}$$

This proves (4.5.7) under the additional condition (4.5.8).

Finally, we show that, for (4.5.7) to hold, condition (4.5.8) can be relaxed to (4.5.2). To this end, we define

$$\widetilde{\theta}(x) := \frac{\theta(x)}{\left(\sum_{j=0}^{\infty} |\theta(2^{-j}x)|^2\right)^{\frac{1}{2}}}.$$

It is obvious that $\widetilde{\theta} \in C^m[0,\infty)$, $\operatorname{supp} \widetilde{\theta} \subset (a,b) \subset (0,\infty)$, and

$$\sum_{j=0}^{\infty} \widetilde{\theta}(2^{-j}x) = 1, \quad \forall x > 0.$$

Thus, using the already proven case of the inequality (4.5.7), we have

$$\|f\|_{\kappa,p} \le c_p \left\| \left(\sum_{j=1}^{\infty} (\Delta_{\widetilde{\theta},j}^\kappa f)^2 \right)^{1/2} \right\|_{\kappa,p}. \tag{4.5.10}$$

Next, let $\phi \in C^\infty[0,\infty)$ be such that $\phi(x) = 1$ for $x \in [a,b]$, and $\operatorname{supp}\phi \subset (a_1,b_1)$ for some $0 < a_1 < a < b < b_1 < \infty$. Define

$$\psi(x) = \frac{\phi(x)}{\left(\sum_{j=0}^\infty |\theta(2^{-j}x)|^2\right)^{\frac{1}{2}}}.$$

Then $\widetilde{\theta}(x) = \theta(x)\psi(x)$, and hence $\Delta^\kappa_{\widetilde{\theta},j} = \Delta^\kappa_{\psi,j} \circ \Delta^\kappa_{\theta,j}$. Thus, we may rewrite (4.5.10) in the form

$$\|f\|_{\kappa,p} \le c_p \left\| \left(\sum_{j=0}^\infty |\Delta^\kappa_{\psi,j} g_j|^2 \right)^{\frac{1}{2}} \right\|_{\kappa,p}, \tag{4.5.11}$$

with $g_j := \Delta^\kappa_{\theta,j} f$. On the other hand, since $\psi \in C^m[0,\infty)$ has compact support, and $m-1 > \lambda_\kappa$, using Theorem 3.5.4 and summation by parts finitely many times, it follows that

$$\sup_{j\in\mathbb{N}} |\Delta^\kappa_{\psi,j} g(x)| \le c_\kappa S_*^{\lambda_\kappa+1}(g)(x) \le C\mathscr{M}_\kappa g(x) + C\mathscr{M}_\kappa g(-x).$$

Thus, by Theorem 4.1.1,

$$\left\| \left(\sum_{j=0}^\infty |\Delta^\kappa_{\psi,j} g_j|^2 \right)^{\frac{1}{2}} \right\|_{\kappa,p} \le c_p \left\| \left(\sum_{j=0}^\infty |g_j|^2 \right)^{\frac{1}{2}} \right\|_{\kappa,p} = c_p \left\| \left(\sum_{j=0}^\infty |\Delta^\kappa_{\theta,j} f|^2 \right)^{\frac{1}{2}} \right\|_{\kappa,p},$$

which shows the desired inverse inequality. □

4.6 Notes and further results

A general Littlewood–Paley theory for a symmetric diffusion semi-group was developed by E. M. Stein in his 1970 monograph [49]. A vector-valued extension of this general theory was developed in [33]. The main references for Section 4.2 are [3, 33, 49].

The multiplier theorem (Theorem 4.4.2) and its analogue on the unit ball and the simplex were proved in [12]. The Littlewood–Paley theory in Section 4.3 and the proof of Theorem 4.4.2 follow the argument in Bonami and Clerc [3], who established the theory in the unweighted setting.

Applications of the refined Littlewood–Paley inequality, Theorem 4.5.2, can be found in [7, 9].

Chapter 5

Sharp Jackson and Sharp Marchaud Inequalities

The goal of this chapter is to prove two inequalities, the sharp Jackson inequality and the sharp Marchaud inequality, for the h-harmonic expansions on the sphere $\mathbb{S}^{d-1}$, which are useful in the embedding theory of function spaces. The multiplier theorem and the Littlewood–Paley inequality established in the prior chapter play crucial roles in their proofs.

As a motivation, these inequalities for trigonometric polynomial approximation on the circle are stated in Section 5.1. Section 5.2 contains several useful properties and results on weighted moduli of smoothness, including the direct Jackson inequality and its inverse. A weighted K-functional that is equivalent with the weighted modulus of smoothness is defined in Section 5.3, using fractional powers of the h-Laplace–Beltrami operator. Weighted sharp Marchaud and sharp Jackson inequalities are proved in Sections 5.4 and 5.5, respectively. Finally, the optimality of the parameters in the sharp Marchaud inequality and the sharp Jackson inequality is proven in Section 5.6.

5.1 Introduction

For trigonometric polynomial approximation of functions on the unit circle $\mathbb{S}^1$ (identified with $[-\pi,\pi]$), M. Timan [56] proved that, for $1<p<\infty$,

$$n^{-r}\Big\{\sum_{k=1}^{n} k^{sr-1}E_k(f)_p^s\Big\}^{1/s} \le C(r,p)\omega^r(f,n^{-1})_p, \quad s=\max(p,2), \tag{5.1.1}$$

where $r\in\mathbb{N}$, $E_k(f)_p$ is the best approximation of $f\in L^p(\mathbb{S}^1)$ by trigonometric polynomials of degree at most k,

$$E_k(f)_p = \min\Big\{\|f-T_n\|_{L_p(\mathbb{S}^1)} : T_n \in \operatorname*{span}_{k<n}\{\sin kt, \cos kt\}\Big\},$$

and $\omega^r(f,t)_p$ denotes the r-th order modulus of smoothness of $f \in L^p(\mathbb{S}^1)$,

$$\omega^r(f,t)_p = \sup_{|h|\le t} \Big\| \sum_{j=0}^{r} (-1)^{r-j} \binom{r}{j} f(\cdot + jh) \Big\|_{L_p(\mathbb{S}^1)}.$$

We call an inequality of the type (5.1.1) a sharp Jackson inequality, since it improves the classical Jackson inequality, $E_n(f)_p \le C\omega^r(f,1/n)_p$, for $1 < p < \infty$.

An estimate of $\omega^r(f,t)_p$ in the direction opposite to (5.1.1) was proved by M. Timan [55] as well: for $1 < p < \infty$ and $r \in \mathbb{N}$,

$$\omega^r(f,1/n)_p \le c(r,p) n^{-r} \Big\{ \sum_{k=1}^{n} k^{rq-1} E_k(f)_p^q \Big\}^{1/q}, \quad q = \min(p,2). \tag{5.1.2}$$

This estimate is, in fact, equivalent to the following inequality on moduli of smoothness: for $1 < p < \infty$ and $r \in \mathbb{N}$,

$$\omega^r(f,t)_p \le \tilde{c}(r,p) t^r \Big\{ \int_t^{1/2} \frac{\omega^{r+1}(f,u)_p^q}{u^{qr+1}} du \Big\}^{1/q}, \quad q = \min(p,2). \tag{5.1.3}$$

The inequality (5.1.3) is stronger than the usual Marchaud inequality

$$\omega^r(f,t)_p \le c(r,p) t^r \int_t^{1/2} \frac{\omega^{r+1}(f,u)_p}{u^{r+1}} du,$$

and accordingly is sometimes called the sharp Marchaud inequality.

Similar to the sharp Marchaud inequality (5.1.3), one has in the other direction the following inequality on moduli of smoothness, which is equivalent to the sharp Jackson inequality (5.1.1): for $1 < p < \infty$,

$$t^r \Big\{ \int_t^{1/2} \frac{\omega^{r+1}(f,u)_p^s}{u^{sr+1}} du \Big\}^{1/s} \le C\omega^r(f,t)_p, \quad s = \max(p,2). \tag{5.1.4}$$

Of particular interest is the case $p = 2$, for which (5.1.1) combined with (5.1.2) yields

$$\omega^r(f,1/n)_2 \sim n^{-r} \Big\{ \sum_{k=1}^{n} k^{2r-1} E_k(f)_2^2 \Big\}^{1/2}.$$

5.2 Moduli of smoothness and best approximation

Let G be a finite reflection group and let h_κ be the weight function defined in (2.1.2), which is invariant under G. Then h_κ is a homogeneous function of degree $\sum_{v\in R_+} \kappa_v$. Recall that

$$\lambda_\kappa = \sum_{v\in R_+} \kappa_v + \frac{d-2}{2}.$$

Associated with the weight h_κ^2, a generalized translation operator T_θ^κ is defined in (3.5.1) for all $\theta \in \mathbb{R}$, which we write as

$$\operatorname{proj}_n^\kappa(T_\theta^\kappa f) = \frac{C_n^{\lambda_\kappa}(\cos\theta)}{C_n^{\lambda_\kappa}(1)} \operatorname{proj}_n^\kappa f, \quad n = 0, 1, \ldots.$$

The operators T_θ^κ are uniformly bounded on $L^p(h_\kappa^2; \mathbb{S}^{d-1})$, as shown in (3.5.3),

$$\|T_\theta^\kappa f\|_{\kappa,p} \le \|f\|_{\kappa,p}, \quad 1 \le p \le \infty.$$

When $\kappa = 0$, T_θ^0 is the usual translation operator on $\mathbb{S}^{d-1}$, aka spherical mean operator.

For $r > 0$ and $0 < \theta < \pi$, we define the r-th order difference operator

$$(I - T_\theta^\kappa)^{r/2} := \sum_{n=0}^{\infty} (-1)^n \binom{r/2}{n} (T_\theta^\kappa)^n,$$

in a distributional sense, by

$$\operatorname{proj}_n^\kappa \left[(I - T_\theta^\kappa)^{r/2} f \right] = \left(1 - \frac{C_n^{\lambda_\kappa}(\cos\theta)}{C_n^{\lambda_\kappa}(1)}\right)^{r/2} \operatorname{proj}_n^\kappa f, \quad n = 0, 1, 2, \ldots.$$

Definition 5.2.1. Let $r > 0$ and $0 < \theta < \pi$. For $f \in L^p(h_\kappa^2; \mathbb{S}^{d-1})$ and $1 \le p < \infty$, or $f \in C(\mathbb{S}^{d-1})$ and $p = \infty$, the weighted r-th order modulus of smoothness is defined as

$$\omega_r(f, t)_{\kappa,p} := \sup_{0<\theta\le t} \|(I - T_\theta^\kappa)^{r/2} f\|_{\kappa,p}, \quad 0 < t < \pi. \tag{5.2.1}$$

This definition makes sense, since the next proposition shows that $\omega_r(f,t)_{\kappa,p}$ satisfies the basic properties of the usual moduli:

Proposition 5.2.2. *Let $f \in L^p(h_\kappa^2)$ if $1 \le p < \infty$ and $f \in C(\mathbb{S}^{d-1})$ if $p = \infty$. Then*

1. $\omega_r(f,t)_{\kappa,p} \le 2^{r+2} \|f\|_{\kappa,p}$;
2. $\omega_r(f,t)_{\kappa,p} \to 0$ *if* $t \to 0^+$;
3. $\omega_r(f,t)_{\kappa,p}$ *is monotone nondecreasing on* $(0,\pi)$;
4. $\omega_r(f+g,t)_{\kappa,p} \le \omega_r(f,t)_{\kappa,p} + \omega_r(g,t)_{\kappa,p}$;
5. *for* $0 < s < r$,
$$\omega_r(f,t)_{\kappa,p} \le 2^{(r-s)+2} \omega_s(f,t)_{\kappa,p}.$$

The proof of Proposition 5.2.2 can be found in [16, Chapter 10] and [72].

Definition 5.2.3. For $f \in L^p(h_\kappa^2)$ and $1 \le p < \infty$, or $f \in C(\mathbb{S}^{d-1})$ and $p = \infty$, the weighted best approximation of f by spherical polynomials of degree at most n is defined as

$$E_n(f)_{\kappa,p} := \inf_{P \in \Pi_{n-1}(\mathbb{S}^{d-1})} \|f - P\|_{\kappa,p}.$$

We will need the near best approximation operator defined via a cut-off function for h-spherical harmonic expansions.

Definition 5.2.4. Let η be a C^∞ function on $[0,\infty)$ such that $\eta(t)=1$ for $0\le t\le 1$ and $\eta(t)=0$ for $t\ge 2$. For $f\in L(h_\kappa^2)$, we define

$$L_n^\kappa f := \sum_{j=0}^{2n} \eta\left(\frac{j}{n}\right) \operatorname{proj}_j^\kappa f, \quad n=1,2,\dots. \tag{5.2.2}$$

The following proposition collects several useful results on the operator L_n^κ.

Proposition 5.2.5. *Let $f\in L^p(h_\kappa^2)$ if $1\le p<\infty$ and $f\in C(\mathbb{S}^{d-1})$ if $p=\infty$. Then*

(1) *$L_n^\kappa f\in \Pi_{2n-1}(\mathbb{S}^{d-1})$ and $L_n^\kappa f = f$ for $f\in\Pi_n(\mathbb{S}^{d-1})$;*

(2) *for $n\in\mathbb{N}$, $\|L_n^\kappa f\|_{\kappa,p}\le c\|f\|_{\kappa,p}$;*

(3) *for $n\in\mathbb{N}$, $\|f-L_n^\kappa f\|_{\kappa,p}\le (1+c)E_n(f)_{\kappa,p}$.*

Theorem 5.2.6. *Let $f\in L^p(h_\kappa^2;\mathbb{S}^{d-1})$ if $1\le p<\infty$, and $f\in C(\mathbb{S}^{d-1})$ if $p=\infty$. Then for any $r>0$ and $n\in\mathbb{N}$,*

$$E_n(f)_{\kappa,p}\le c\,\omega_r(f,n^{-1})_{\kappa,p}, \tag{5.2.3}$$

and

$$\omega_r(f,n^{-1})_{\kappa,p}\le cn^{-r}\sum_{k=0}^{n}(k+1)^{r-1}E_k(f)_{\kappa,p}. \tag{5.2.4}$$

The proofs of Proposition 5.2.5 and Theorem 5.2.6 can be found in [16, Chapter 10].

5.3 Weighted Sobolev spaces and K-functionals

Recall that the space $\mathscr{H}_n^d(h_\kappa^2)$ of h-spherical harmonics on $\mathbb{S}^{d-1}$ of degree n is an eigenvector space of the h-Laplace–Beltrami operator $\Delta_{h,0}$; namely,

$$\mathscr{H}_n^d(h_\kappa^2)=\left\{f\in C^2(\mathbb{S}^{d-1}):\ \Delta_{h,0}f=-n(n+2\lambda_\kappa)f\right\},\quad n=0,1,\dots.$$

Accordingly, we can define fractional powers of $\Delta_{h,0}$ and the weighted Sobolev spaces as follows.

Definition 5.3.1. For $r>0$ and $1\le p\le\infty$, a function $f\in L^p(h_\kappa^2;\mathbb{S}^{d-1})$ is said to belong to the weighted Sobolev space $\mathscr{W}_p^r(h_\kappa^2)$ if there exists a function $g\in L^p(h_\kappa^2;\mathbb{S}^{d-1})$, which we denote by $(-\Delta_{h,0})^{r/2}f$, such that

$$\operatorname{proj}_n^\kappa\left[(-\Delta_{h,0})^{r/2}f\right]=(n(n+2\lambda_\kappa))^{r/2}\operatorname{proj}_n^\kappa f,\quad n=0,1,\dots, \tag{5.3.1}$$

where we assume $f,g \in C(\mathbb{S}^{d-1})$ when $p=\infty$. The norm in the weighted Sobolev space $\mathscr{W}_p^r(h_\kappa^2)$ is defined by

$$\|f\|_{\mathscr{W}_p^r(h_\kappa^2)} := \|f\|_{\kappa,p} + \|(-\Delta_{h,0})^{r/2} f\|_{\kappa,p}.$$

The fractional spherical h-Laplacian $(-\Delta_{h,0})^{r/2}$ is then a linear operator on the space $\mathscr{W}_p^r(h_\kappa^2)$ defined by (5.3.1).

Let $\eta \in C^\infty[0,\infty)$ be as in Definition 5.2.4, and let $\theta(x) = \eta(2x) - \eta(4x)$. Let $\triangle_{\theta,j}^\kappa$, $j=0,1,\ldots$, be the operators defined in Definition 4.5.1. Obviously, $\triangle_{\theta,j}^\kappa(f) := L_{2^j-1}^\kappa f - L_{2^{j-2}}^\kappa f$ and, for $f \in L^p(h_\kappa^2;\mathbb{S}^{d-1})$ if $1 \le p < \infty$ or $f \in C(\mathbb{S}^{d-1})$ if $p=\infty$,

$$f = \sum_{j=0}^{\infty} \triangle_{\theta,j}^\kappa f, \tag{5.3.2}$$

where the series converges in the norm of $L^p(h_\kappa^2)$.

A Littlewood–Paley type inequality holds on the weighted Sobolev spaces $\mathscr{W}_p^r(h_\kappa^2)$.

Theorem 5.3.2. *For $1 < p < \infty$, $\gamma \ge 0$, and $f \in W_p^\gamma(h_\kappa^2)$,*

$$\left\| \left\{ \sum_{j=1}^{\infty} 2^{2j\gamma} \left(\triangle_{\theta,j}^\kappa f\right)^2 \right\}^{1/2} \right\|_{\kappa,p} \sim \|(-\Delta_{h,0})^{\gamma/2} f\|_{\kappa,p}, \tag{5.3.3}$$

and

$$\left\| \left\{ \sum_{j=0}^{\infty} 2^{2j\gamma} \left(\triangle_{\theta,j}^\kappa f\right)^2 \right\}^{1/2} \right\|_{\kappa,p} \sim \|f\|_{\mathscr{W}_p^\gamma(h_\kappa^2)}, \tag{5.3.4}$$

where the constants of equivalence depend only on p, d and γ.

Theorem 5.3.2 is a consequence of the Marcinkiewicz multiplier theorem, Theorem 4.4.2, and its proof runs along the same line as that of Theorem 4.5.2.

For $f \in L^p(h_\kappa^2)$, we define its K-functional in terms of the h-spherical Laplacian as follows.

Definition 5.3.3. Given $r > 0$, the r-th K-functional of $f \in L^p(h_\kappa^2)$ is

$$K_r(f;t)_{\kappa,p} := \inf_{g \in \mathscr{W}_p^r(h_\kappa^2)} \left\{ \|f-g\|_{\kappa,p} + t^r \left\| (-\Delta_{h,0})^{r/2} g \right\|_{\kappa,p} \right\}. \tag{5.3.5}$$

We have the following realization theorem of the K-functional, whose proof can be found in [16, Chapter 10] and [21].

Theorem 5.3.4. *Let $f \in L^p(h_\kappa^2)$ if $1 \le p < \infty$ and $f \in C(\mathbb{S}^{d-1})$ if $p=\infty$. If $t \in (0,1)$ and n is a positive integer such that $n \sim t^{-1}$, then*

$$K_r(f,t)_{\kappa,p} \sim \|f - L_n^\kappa f\|_{\kappa,p} + n^{-r} \left\| (-\Delta_{h,0})^{r/2} L_n^\kappa f \right\|_{\kappa,p}. \tag{5.3.6}$$

It turns out that the weighted modulus of smoothness $\omega_r(f,t)_{\kappa,p}$ and the weighted K-functional $K_r(f;t)_{\kappa,p}$ are equivalent.

Theorem 5.3.5. *Let $f \in L^p(h_\kappa^2)$ if $1 \le p < \infty$ and let $f \in C(\mathbb{S}^{d-1})$ if $p = \infty$. If $t \in (0, \frac{\pi}{2})$ and $r > 0$, then*

$$\omega_r(f,t)_{\kappa,p} \sim \left\| (I - T_t^\kappa)^{r/2} f \right\|_{\kappa,p} \sim K_r(f,t)_{\kappa,p}.$$

The proof of Theorem 5.3.5 can be found in [16, Chapter 10]. Theorem 5.3.5 combined with Theorem 5.2.6 yields the following direct Jackson inequality and the inverse inequality.

Theorem 5.3.6. *Let $f \in L^p(h_\kappa^2; \mathbb{S}^{d-1})$ if $1 \le p < \infty$, and $f \in C(\mathbb{S}^{d-1})$ if $p = \infty$. Then for any $r > 0$ and $n \in \mathbb{N}$,*

$$E_n(f)_{\kappa,p} \le c K_r(f; n^{-1})_{\kappa,p},$$

and

$$K_r(f, n^{-1})_{\kappa,p} \le c n^{-r} \sum_{k=0}^{n} (k+1)^{r-1} E_k(f)_{\kappa,p}.$$

5.4 The sharp Marchaud inequality

In this section, we will prove the following sharp Marchaud inequality.

Theorem 5.4.1. *For $\alpha > 0$, $1 < p < \infty$ and $q = \min(p,2)$,*

$$\omega_\alpha(f,t)_{\kappa,p} \le ct^\alpha \left(\int_t^1 \frac{\omega_{\alpha+1}(f,u)_{\kappa,p}^q}{u^{q\alpha+1}} du \right)^{\frac{1}{q}}. \tag{5.4.1}$$

Using Theorem 5.2.6, it can be easily seen that Theorem 5.4.1 is, in fact, equivalent to the following sharp inverse inequality.

Corollary 5.4.2. *For $1 < p < \infty$, $q = \min(p,2)$ and $\alpha > 0$,*

$$\omega_\alpha(f, 1/n)_{\kappa,p} \le c(\alpha,p) n^{-\alpha} \Big(\sum_{j=1}^{n} j^{\alpha q - 1} E_j(f)_{\kappa,p}^q \Big)^{1/q}. \tag{5.4.2}$$

The proof of Theorem 5.4.1 relies only on the Cesàro summability of the orthogonal expansions, and on Theorem 4.2.4 (the Littlewood–Paley–Stein theorem).

Proof of Theorem 5.4.1. By Theorem 5.3.5 and (5.3.6), it suffices to show that

$$2^{-m\alpha} \|(-\Delta_{h,0})^{\alpha/2} (L_{2^m}^\kappa(f))\|_{\kappa,p} \le c 2^{-m\alpha} \left(\sum_{j=0}^{m} 2^{j\alpha q} K_{\alpha+1}(f, 2^{-j})_{\kappa,p}^q \right)^{\frac{1}{q}}, \tag{5.4.3}$$

where $q = \min\{p,2\}$. For convenience, we set $F = (-\Delta_{h,0})^{\alpha/2}(L_{2^m}^\kappa(f))$, and use $S_n^\delta(f)$ to denote the Cesàro (C,δ)-means $S_n^\delta(h_\kappa^2; f)$ defined in (3.3.7).

For the proof of (5.4.3) we claim that for, $\delta \ge 0$,

$$g(F) \le c\left(\sum_{j=0}^{\infty} 2^{-j(1+q)} \sum_{i=2^j}^{2^{j+1}-1} |S_i^\delta((-\Delta_{h,0})^{1/2}F)|^q\right)^{\frac{1}{q}}, \tag{5.4.4}$$

where $g(f)$ denotes the the Littlewood–Paley–Stein g-function defined in (4.2.2).

For the moment, we take the claim (5.4.4) for granted and proceed with the proof. Using (5.4.4) and Theorem 4.2.4, we obtain

$$\|F\|_{\kappa,p} \le C\|g(F)\|_{\kappa,p} \le c\left(\sum_{j=0}^{\infty} 2^{-j(1+q)} \sum_{i=2^j}^{2^{j+1}-1} \|S_i^\delta((-\Delta_{h,0})^{1/2}F)\|_{\kappa,p}^q\right)^{\frac{1}{q}}, \tag{5.4.5}$$

where the last step uses the Minkowski inequality for $p > 2$. We break the first sum on the right side of (5.4.5) into two parts: $\sum_{j=0}^{m-4}$ and $\sum_{j=m-3}^{\infty}$. Observe that if $2^j \le i \le 2^{j+1}-1$ and $0 \le j \le m-4$, then

$$\begin{aligned} S_i^\delta((-\Delta_{h,0})^{1/2}F) &= L_{2^{j+2}}^\kappa(S_i^\delta((-\Delta_{h,0})^{1/2}F)) = S_i^\delta(L_{2^{j+2}}^\kappa((-\Delta_{h,0})^{1/2}F)) \\ &= S_i^\delta((-\Delta_{h,0})^{(\alpha+1)/2}(L_{2^{j+2}}^\kappa(f))). \end{aligned} \tag{5.4.6}$$

Thus, if $\delta > \lambda_\kappa$,

$$\begin{aligned} &\left(\sum_{j=0}^{m-4} 2^{-j(1+q)} \sum_{i=2^j}^{2^{j+1}-1} \|S_i^\delta((-\Delta_{h,0})^{1/2}F)\|_{\kappa,p}^q\right)^{\frac{1}{q}} \\ &\qquad \le c\left(\sum_{j=0}^{m-4} 2^{-jq}\left\|(-\Delta_{h,0})^{(\alpha+1)/2}(L_{2^{j+2}}^\kappa(f))\right\|_{\kappa,p}^q\right)^{\frac{1}{q}} \\ &\qquad \le c\left(\sum_{j=0}^{m-4} 2^{j\alpha q} K_{\alpha+1}(f,2^{-j})_{\kappa,p}^q\right)^{\frac{1}{q}}, \end{aligned} \tag{5.4.7}$$

where the second step uses (5.4.6) and the Cesàro (C,δ)-summability of h-harmonic expansions for $\delta > \lambda_\kappa$, and the last step uses (5.3.6). However, on the other hand, for $\delta > \lambda_\kappa$,

$$\begin{aligned} &\left(\sum_{j=m-3}^{\infty} 2^{-j(1+q)} \sum_{i=2^j}^{2^{j+1}-1} \|S_i^\delta((-\Delta_{h,0})^{1/2}F)\|_{\kappa,p}^q\right)^{\frac{1}{q}} \le c\left(\sum_{j=m-3}^{\infty} 2^{-jq}\right)^{\frac{1}{q}} \|(-\Delta_{h,0})^{1/2}F\|_{\kappa,p} \\ &\qquad \le c2^{-m}\|(-\Delta_{h,0})^{1/2}F\|_{\kappa,p} = c2^{-m}\|(-\Delta_{h,0})^{(\alpha+1)/2}(L_{2^m}^\kappa(f))\|_{\kappa,p} \\ &\qquad \le c2^{m\alpha}K_{\alpha+1}(f,2^{-m})_{\kappa,p}, \end{aligned} \tag{5.4.8}$$

where the last step uses the realization (5.3.6). Therefore, combining (5.4.5), (5.4.7), and (5.4.8), we deduce the desired estimate (5.4.3).

It remains to prove claim (5.4.4). Note that

$$\frac{\partial}{\partial r}P_r^\kappa(F)=\sum_{i=1}^{\infty} ir^{i-1}\operatorname{proj}_i^\kappa(F)=(1-r)^{\delta+1}\sum_{i=1}^{\infty}\left(\sum_{j=1}^{i} jA_{i-j}^\delta\operatorname{proj}_j^\kappa(F)\right)r^{i-1},$$

where we used the identity $\sum_{k=0}^\infty A_k^\delta r^k=(1-r)^{-1-\delta}$ in the last step. Thus,

$$\begin{aligned}\left|\frac{\partial}{\partial r}P_r^\kappa(F)\right| &\le (1-r)^{\delta+1}\sum_{i=1}^{\infty} A_i^\delta |S_i^\delta((-\Delta_{h,0})^{1/2}F)|r^{i-1}\\ &\le c(1-r)^{1+\delta}\sum_{j=0}^{\infty}2^{j\delta}r^{2^j-1}\sum_{i=2^j}^{2^{j+1}-1}|S_i^\delta((-\Delta_{h,0})^{1/2}F)|,\end{aligned}$$

which, using the Cauchy–Schwartz inequality, implies

$$\begin{aligned}\left|\frac{\partial}{\partial r}P_r^\kappa(F)\right|^2 &\le c(1-r)^{2+2\delta}\Big(\sum_{j=0}^{\infty}2^{j\delta}r^{2^j-1}\Big(\sum_{i=2^j}^{2^{j+1}-1}|S_i^\delta((-\Delta_{h,0})^{1/2}F)|\Big)^2\Big)\Big(\sum_{\ell=0}^{\infty}2^{\ell\delta}r^{2^\ell-1}\Big)\\ &= c\sum_{j=0}^{\infty}\Big[2^{j\delta}\Big(\sum_{i=2^j}^{2^{j+1}-1}|S_i^\delta((-\Delta_{h,0})^{1/2}F)|\Big)^2\sum_{\ell=0}^{\infty}2^{\ell\delta}r^{2^\ell+2^j-2}(1-r)^{2+2\delta}\Big].\end{aligned}\tag{5.4.9}$$

On the other hand, a straightforward calculation shows that

$$\begin{aligned}&\sum_{\ell=0}^{\infty}2^{\ell\delta}\int_0^1 r^{2^\ell+2^j-2}(1-r)^{2+2\delta}r|\log r|\,dr\le c\sum_{\ell=0}^{\infty}2^{\ell\delta}\int_0^1 r^{2^\ell+2^j-2}(1-r)^{3+2\delta}\,dr\\ &\quad= c\sum_{\ell=0}^{\infty}2^{\ell\delta}\frac{\Gamma(2^\ell+2^j-1)\Gamma(4+2\delta)}{\Gamma(2^\ell+2^j+3+2\delta)}\le c\sum_{\ell=0}^{\infty}2^{\ell\delta}(2^\ell+2^j)^{-4-2\delta}\\ &\quad\le c2^{-j(\delta+4)}.\end{aligned}$$

Thus, using (5.4.9) and (4.2.2), we conclude that

$$g(F)\le c\Big(\sum_{j=0}^{\infty}2^{-4j}\Big(\sum_{i=2^j}^{2^{j+1}-1}|S_i^\delta((-\Delta_{h,0})^{1/2}F)|\Big)^2\Big)^{\frac12},\tag{5.4.10}$$

which implies claim (5.4.4) for $p>2$.

Finally, for $1<p\le 2$, we use (5.4.10) and Hölder's inequality to obtain

$$\begin{aligned}|g(F)|^p &\le c\sum_{j=0}^{\infty}2^{-2jp}\left(\sum_{i=2^j}^{2^{j+1}-1}|S_i^\delta((-\Delta_{h,0})^{1/2}F)|\right)^p\\ &\le c\sum_{j=0}^{\infty}2^{-j(p+1)}\sum_{i=2^j}^{2^{j+1}-1}|S_i^\delta((-\Delta_{h,0})^{1/2}F)|^p,\end{aligned}$$

which proves (5.4.4) for $1<p\le 2$ as well. □

5.5 The sharp Jackson inequality

The main goal in this section is to prove the following sharp Jackson inequality.

Theorem 5.5.1. *For $f \in L^p(h_\kappa^2;\mathbb{S}^{d-1})$, $1 < p < \infty$, $r > 0$, and $s = \max(p,2)$,*

$$t^r\Big\{\sum_{1\le j\le 1/t} j^{sr-1}E_j(f)_{\kappa,p}^s\Big\}^{1/s} \le c\omega^r(f,t)_{\kappa,p}. \tag{5.5.1}$$

Using Hardy's inequality and Theorem 5.2.6, it is easily seen that Theorem 5.5.1 is, in fact, equivalent to the following corollary:

Corollary 5.5.2. *For $f \in L^p(h_\kappa^2;\mathbb{S}^{d-1})$, $1 < p < \infty$, and $s = \max(p,2)$,*

$$t^r\Big\{\int_t^{1/2}\frac{\omega^{r+1}(f,u)_{\kappa,p}^s}{u^{rs+1}}\,du\Big\}^{1/s} \le c\omega^r(f,t)_{\kappa,p}.$$

Proof of Theorem 5.5.1. Obviously, it suffices to show that

$$2^{-nr}\Big(\sum_{j=1}^n 2^{jrs}E_{2^j}(f)_{\kappa,p}^s\Big)^{1/s} \le cK_r(f,2^{-n})_{\kappa,p}. \tag{5.5.2}$$

Setting $g_n = L_{2^{n-1}}^\kappa f$, we have

$$E_{2^n}(f)_{\kappa,p} \le \|f-g_n\|_{\kappa,p} \le cK_r\big(f,2^{-nr}\big)_{\kappa,p}.$$

As $E_m(f-g_n)_{\kappa,p} \le \|f-g_n\|_{\kappa,p}$ for all $m \in \mathbb{N}$, we obtain

$$E_{2^j}(f)_{\kappa,p} \le E_{2^j}(f-g_n)_{\kappa,p} + E_{2^j}(g_n)_{\kappa,p} \le \|f-g_n\|_{\kappa,p} + E_{2^j}(g_n)_{\kappa,p}\,.$$

We can now write

$$\begin{aligned}
&2^{-nr}\Big(\sum_{j=1}^n 2^{jrs}E_{2^j}(f)_{\kappa,p}^s\Big)^{1/s}\\
&\quad\le 2^{-nr}\Big(\sum_{j=1}^n 2^{jrs}E_{2^j}(f-g_n)_{\kappa,p}^s\Big)^{1/s} + 2^{-nr}\Big(\sum_{j=1}^n 2^{jrs}E_{2^j}(g_n)_{\kappa,p}^s\Big)^{1/s}\\
&\quad\le \frac{2^r}{(2^{rs}-1)^{1/s}}\|f-g_n\|_{\kappa,p} + 2^{-nr}\Big(\sum_{j=1}^n 2^{jrs}E_{2^j}(g_n)_{\kappa,p}^s\Big)^{1/s}.
\end{aligned}$$

Therefore, for the proof of (5.5.2), it remains to show that

$$2^{-nr}\Big(\sum_{j=1}^n 2^{jrs}E_{2^j}(g_n)_{\kappa,p}^s\Big)^{1/s} \le cK_r\big(f,2^{-n}\big)_{\kappa,p},$$

which, using (5.3.6), is a direct consequence of the inequality

$$\sum_{j=1}^{n} 2^{jrs} E_{2^j}(g_n)_{\kappa,p}^s \le c\|(-\Delta_{h,0})^{r/2} g_n\|_{\kappa,p}^s. \tag{5.5.3}$$

For the proof of (5.5.3), we write for $j \le n$,

$$E_{2^{j+1}}(g_n)_{\kappa,p} \le \|g_n - L_{2^j} g_n\|_{\kappa,p} = \|L_{2^n} g_n - L_{2^j} g_n\|_{\kappa,p} = \Big\| \sum_{\ell=j+2}^{n+1} \triangle_{\theta,\ell}^{\kappa} g_n \Big\|_{\kappa,p}.$$

Note that $L_{2^i}^{\kappa}(L_{2^n}^{\kappa} f - L_{2^j}^{\kappa} f) = 0$ for $i < j \le n$, and $\triangle_{\theta,i}^{\kappa}(L_{2^n}^{\kappa} g_n - L_{2^j}^{\kappa} g_n) = 0$ for $i > n+1$. Thus, applying Theorem 5.3.2 to the function $L_{2^n}^{\kappa} g_n - L_{2^j}^{\kappa} g_n$, we deduce that

$$\|L_{2^n}^{\kappa} g_n - L_{2^j}^{\kappa} g_n\|_{\kappa,p} \sim \Big\| \Big(\sum_{\ell=j+1}^{n+1} (\triangle_{\theta,\ell}^{\kappa} g_n)^2 \Big)^{1/2} \Big\|_{\kappa,p}.$$

Thus, proving (5.5.3) reduces to showing that

$$\sum_{j=0}^{n+1} 2^{jrs} \Big\| \Big(\sum_{\ell=j+1}^{n+1} (\triangle_{\theta,\ell}^{\kappa} g_n)^2 \Big)^{1/2} \Big\|_{\kappa,p}^s \le c\|(-\Delta_{h,0})^{r/2} g_n\|_{\kappa,p}^s. \tag{5.5.4}$$

We prove (5.5.4) separately for $1 < p \le 2$, in which case $s = 2$, and for $2 < p < \infty$, in which case $s = p$. For $1 < p \le 2$ we use $\|f\|_q + \|g\|_q \le \| \, |f| + |g| \, \|_q$ for the quasi-norm $\|\cdot\|_q$ when $q \le 1$, and obtain

$$\begin{aligned} \sum_{j=1}^{n+1} 2^{jr2} \Big\| \sum_{\ell=j+1}^{n+1} (\triangle_{\theta,\ell}^{\kappa} g_n)^2 \Big\|_{p/2} &\le \Big\| \sum_{j=1}^{n+1} 2^{jr2} \sum_{\ell=j+1}^{n+1} (\triangle_{\theta,\ell}^{\kappa} g_n)^2 \Big\|_{p/2} \\ &\le c \Big\| \Big(\sum_{\ell=2}^{n+1} (\triangle_{\theta,\ell}^{\kappa} g_n)^2 2^{\ell r2} \Big)^{1/2} \Big\|_{\kappa,p}^2 \le c\|(-\Delta_{h,0})^{r/2} g_n\|_{\kappa,p}^s, \end{aligned}$$

where the last step uses Theorem 5.3.2. This proves (5.5.4) for the case $1 < p \le 2$.

Finally, we prove (5.5.4) for the case of $2 < p < \infty$. Setting $E := \{(j,\ell) : j,\ell \in \mathbb{Z}, \, 0 \le j \le \ell - 1 \le n\}$, we have

$$\begin{aligned} \int_{\mathbb{S}^{d-1}} \sum_{j=0}^{n+1} 2^{jrp} \Big(\sum_{\ell=j+1}^{n+1} (\triangle_{\theta,\ell}^{\kappa} g_n)^2 \Big)^{p/2} &= \int_{\mathbb{S}^{d-1}} \Big(\sum_{j=0}^{n+1} 2^{jrp} \Big| \sum_{\ell=1}^{n+1} \chi_E(j,\ell) |\triangle_{\theta,\ell}^{\kappa} g_n|^2 \Big|^{p/2} \Big)^{\frac{2}{p}\frac{p}{2}} \\ \le \int_{\mathbb{S}^{d-1}} \Big(\sum_{\ell=1}^{n+1} |\triangle_{\theta,\ell}^{\kappa} g_n|^2 \Big(\sum_{j=0}^{n+1} 2^{jrp} \chi_E(j,\ell) \Big)^{\frac{2}{p}} \Big)^{\frac{p}{2}} &\le c \Big\| \Big(\sum_{\ell=2}^{n+1} |\triangle_{\theta,\ell}^{\kappa}(g_n)|^2 2^{\ell r2} \Big)^{1/2} \Big\|_{\kappa,p}^p, \end{aligned}$$

where the second step uses the weighted Minkowskii inequality. This together with (5.3.3) proves (5.5.4) for $p > 2$. □

5.6 Optimality of the power in the Marchaud inequality

In this section, we show the optimality of the powers s and q in the sharp Jackson inequality (5.5.1) and the sharp Marchaud inequality (5.4.1). More precisely, we have

Theorem 5.6.1. *For* $1 < p < \infty$,

$$\max\{p,2\} = \min\left\{ s>0: \ \sup_f \sup_{n\in\mathbb{N}} \frac{n^{-r}\left(\sum_{k=1}^n k^{rs-1}E_k(f)_{\kappa,p}^s\right)^{1/s}}{\omega^r(f,n^{-1})_{\kappa,p}} < \infty \right\}, \tag{5.6.1}$$

$$\min\{p,2\} = \max\left\{ q>0: \ \sup_f \sup_{n} \frac{\omega^r(f,n^{-1})_{\kappa,p}}{n^{-r}\left(\sum_{k=1}^n k^{rq-1}E_k(f)_{\kappa,p}^q\right)^{\frac{1}{q}}} < \infty \right\}, \tag{5.6.2}$$

where the supremums $\sup_f$ *are taken over all functions* $f \in L^p(h_\kappa^2)$ *that are not constant.*

By slight modifications of the examples in [9], we can deduce the optimality (5.6.1) for all $1 < p < \infty$ and the optimality (5.6.2) for $1 < p \le 2$. The main goal in this section is to show that (5.6.2) holds for $2 \le p < \infty$ as well. By Theorem 5.3.5, it is sufficient to construct a sequence of functions f_n such that

$$K_r(f_n, 2^{-n})_{\kappa,p} \sim c2^{-nr}\left(\sum_{k=1}^n 2^{2kr}E_{2^k}(f_n)_{\kappa,p}^2\right)^{1/2}, \quad 2 \le p < \infty \tag{5.6.3}$$

and

$$\lim_{n\to\infty} \frac{K_r(f_n,2^{-n})_{\kappa,p}}{2^{-nr}\left(\sum_{k=1}^n 2^{qkr}E_{2^k}(f_n)_{\kappa,p}^q\right)^{1/q}} = \infty, \quad \forall q > 2. \tag{5.6.4}$$

The construction of the sequence of functions f_n with the above properties relies on the following crucial proposition, whose proof can be found in [11].

Proposition 5.6.2. *Let* X *be a linear subspace of* Π_N^d *with* $\dim X \ge \varepsilon \dim \Pi_N^d$ *for some* $\varepsilon \in (0,1)$. *Then there exists a function* $f \in X$ *such that* $\|f\|_{\kappa,p} \sim 1$ *for all* $0 < p \le \infty$ *with the constants of equivalence depending only on* ε, d, κ *and* p, *when* p *is small.*

Proofs of (5.6.3) *and* (5.6.4). For each $j \in \mathbb{N}$, let

$$X_j := \bigoplus_{2^{j-1}<k\le 2^j} \mathscr{H}_k^d.$$

Since $\dim \mathscr{H}_k^d(h_\kappa^2) \sim k^{d-2}$, it follows that

$$\dim X_j \sim 2^{j(d-1)} \sim \dim \Pi_{2^j}^d.$$

Thus, using Proposition 5.6.2, there exists a spherical polynomial $P_j \in \bigoplus_{2^{j-1}<k\le 2^j} \mathscr{H}_k^d(h_\kappa^2)$ such that $\|P_j\|_\infty \sim \|P_j\|_2 \sim 1$ for each $j \in \mathbb{N}$. Let $f_n = \sum_{j=1}^n 2^{-jr}P_j$. Using (5.3.5), we

obtain

$$K_r(f_n,2^{-n})_{\kappa,p} \ge c2^{-nr}\|(-\Delta_{h,0})^{\frac{r}{2}} f_n\|_{\kappa,p} \ge c2^{-nr}\|\sum_{j=0}^{n} 2^{-jr}(-\Delta_{h,0})^{\frac{r}{2}} P_j\|_2$$
$$= c2^{-nr}\Big(\sum_{j=0}^{n} 2^{-2jr}\|(-\Delta_{h,0})^{\frac{r}{2}} P_j\|_2^2\Big)^{\frac{1}{2}} \sim 2^{-nr}\Big(\sum_{j=0}^{n} \|P_j\|_2^2\Big)^{\frac{1}{2}}$$
$$\sim 2^{-nr}\sqrt{n}.$$

On the other hand,

$$E_{2^j}(f_n)_{\kappa,p} \le \|\sum_{i=j+1}^{n} 2^{-ir}P_i\|_{\kappa,p} \le \sum_{i=j+1}^{n} 2^{-ir}\|P_i\|_{\kappa,p} \le c2^{-jr}.$$

Thus, for any $2 \le q < \infty$,

$$2^{-nr}\Big(\sum_{j=1}^{n} 2^{jqr}E_{2^j}(f_n)_{\kappa,p}^q\Big)^{1/q} \le c2^{-nr}\Big(\sum_{j=1}^{n} 2^{jqr}2^{-jqr}\Big)^{1/q} \le c2^{-nr}n^{1/q}$$
$$\le cn^{\frac{1}{q}-\frac{1}{2}}K_r(f_n,2^{-n})_{\kappa,p},$$

which implies (5.6.4) for $q > 2$, and the lower estimate of (5.6.3) for $q = 2$. Finally, the upper estimate of (5.6.3) follows directly from (5.4.2). □

5.7 Notes and further results

Inequalities (5.1.1) and (5.1.2) were proved by M. Timan [55] and Zygmund [75]. They were generalized in several articles (see [7, 8, 20, 22, 61]) and described in the texts [19, p. 210], [57, p.338 (12)], and [58, (4.88), p. 191]. Other useful inequalities on moduli of smoothness and their applications in embedding theory can be found in [24, 47, 54].

The weighted moduli of smoothness (5.2.1) and K-functionals (5.3.5) were defined and studied in [72], where the direct and inverse theorem, namely Theorem 5.2.6, and the equivalence $\omega_r(f,t)_{\kappa,p} \sim K_r(f,t)_{\kappa,p}$, as well as several other useful properties of $\omega_r(f,t)_{\kappa,p}$ were established; see also [74]. For polynomial approximation on the unweighted sphere, we refer the reader to the book [64].

Most of the results in Section 5.3 for the K-functionals were proved by Ditzian [21] in a more general setting, where only the Cesàro summability was assumed.

The proof of Theorem 5.5.1, the sharp Jackson inequality, follows along the same lines as [9], where the theorem was shown in a more general setting. A very elegant alternative proof of the sharp Jackson inequality was recently discovered in [22] where, instead of the Littlewood–Paley inequality, only semi-groups and convexity properties of L_p-spaces are used. The method in [22] works also for more general Banach spaces.

The proof of the sharp Marchaud inequality, namely Theorem 5.4.1, follows along the same lines as [7] and [8]. An alternative approach to the sharp Marchaud inequalities without using the Littlewood–Paley inequalities can be found in [20].

The proof of Proposition 5.6.2 is from [11], whereas its idea can be traced back to [67].

Chapter 6

Dunkl Transform

The Dunkl transform is a generalization of the Fourier transform and is an isometry in $L^2(\mathbb{R}^d, h_\kappa^2)$ with h_κ being a reflection invariant weight function. In this chapter we study the Dunkl transform from the point of view of harmonic analysis. In Section 6.1 we show that the Dunkl transform is an isometry in L^2 space with respect to the measure $h_\kappa^2(x)dx$ on $\mathbb{R}^d$ and it preserves Schwartz class of functions. The inversion formula when both f and its Dunkl transform are in L^1 is proved in Section 6.2, for which we consider an approximation operator defined by a generalized convolution with the delation of the Gaussian kernel. The convolution structure is defined in terms of a generalized translation operator, which is defined in the Dunkl transform side; this translation operator is studied in Section 6.3 and its boundedness is established in some restricted classes of functions. The boundedness of the generalized convolution operator is studied in Section 6.4 and used to study the summability of the inverse Dunkl transform. Finally, in Section 6.5, we consider analogues of the Hardy–Littlewood maximal functions in the weighted L^p spaces and prove that they are strong type (p,p) and weak type $(1,1)$, which lead to almost everywhere convergence of the summability methods.

6.1 Dunkl transform: L^2 theory

Recall that V_κ is the intertwining operator associated with a reflection group G and a multiplicity function κ. We define

$$E(x,y) := V_\kappa^{(x)} e^{\langle x,y \rangle}, \qquad x,y \in \mathbb{R}^d.$$

Recall that $E_n(x,y) = V_\kappa^{(x)}(\langle x,y\rangle^n/n!)$. Since the function $y \mapsto f_y(x) := e^{\langle x,y\rangle}$ is in $A(\mathbb{B}^d)$ and $\|f_y\|_A = e^{\|y\|}$, the sum $E(x,y) = \sum_{n=0}^{\infty} E_n(x,y)$ converges uniformly and absolutely on compact sets. In particular, by Proposition 2.3.8, E is symmetric, $E(x,y) = E(y,x)$.

The function $E(x,iy)$ plays the role of $e^{i\langle x,y\rangle}$ in the usual case of the Fourier transform. Since V_κ is a positive operator, $|E(x,iy)| \le 1$.

Definition 6.1.1. Let h_κ be defined as in (2.1.2). For $f \in L^1(\mathbb{R}^d, h_\kappa^2)$, the Dunkl transform is defined by

$$\mathscr{F}_\kappa f(y) := b_h \int_{\mathbb{R}^d} f(x) E(x, -iy) h_\kappa^2(x) dx, \quad y \in \mathbb{R}^d,$$

where b_h is the constant

$$b_h = \left(\int_{\mathbb{R}^d} h_\kappa^2(x) e^{-\|x\|^2/2} dx \right)^{-1} = \frac{c_h}{(2\pi)^{d/2}},$$

and c_h is as in (2.1.6).

If $\kappa = 0$, then $V_\kappa = id$ and the Dunkl transform coincides with the usual Fourier transform.

Theorem 6.1.2. *For $f \in L^1(\mathbb{R}^d, h_\kappa^2)$, the Dunkl transform $\mathscr{F}_\kappa f$ is a bounded continuous function, and $|\mathscr{F}_\kappa f(x)| \le \|f\|_{\kappa,1}$, $x \in \mathbb{R}^d$, where $\|\cdot\|_{\kappa,p}$ denotes the norm of $L^p(\mathbb{R}^d; h_\kappa^2)$.*

Proof. Since $|E(x, iy)| \le 1$, the inequality $|\mathscr{F}_\kappa f(x)| \le \|f\|_{\kappa,1}$ follows immediately. The continuity follows from the continuity of $E(x, -iy)$ and the dominated convergence theorem. □

Proposition 6.1.3. *Let $\nu(z) = z_1^2 + \cdots + z_d^2$, $z_i \in \mathbb{C}$. For $y, z \in \mathbb{C}^d$,*

$$b_h \int_{\mathbb{R}^d} E(x, y) E(x, z) h_\kappa^2(x) e^{-\|x\|^2/2} dx = e^{(\nu(y)+\nu(z))/2} E(y, z). \tag{6.1.1}$$

Proof. First we prove that, for p being a polynomial on $\mathbb{R}^d$ and $y \in \mathbb{C}^d$,

$$b_h \int_{\mathbb{R}^d} (e^{-\Delta_h/2} p(x)) E(x, y) h_\kappa^2(x) e^{-\|x\|^2/2} dx = e^{\nu(y)/2} p(y). \tag{6.1.2}$$

Let m be an integer larger than the degree of p, fix $y \in \mathbb{C}^d$, and let $q_m(x) = \sum_{j=0}^m E_j(x, y)$. Decomposing p into homogeneous components shows that $\langle q_m, p \rangle_{\mathscr{D}} = p(y)$. By (3.1.16), then,

$$p(y) = \langle q_m, p \rangle_{\mathscr{D}} = c_h \int_{\mathbb{R}^d} (e^{-\Delta_h/2} p(x)) \left(e^{-\Delta_h/2} q_m(x)\right) h_\kappa^2(x) e^{-\|x\|^2/2} dx.$$

Since $\Delta_h^{(x)} E_n(x, y) = \nu(y) E_{n-2}(x, y)$, it follows that

$$\begin{aligned} e^{-\Delta_h/2} q_m(x) &= \sum_{j=0}^{m} \sum_{l \le j/2} \frac{1}{l!} \left(\frac{-\nu(y)}{2} \right)^l E_{j-2l}(x, y) \\ &= \sum_{l \le m/2} \frac{1}{l!} \left(\frac{-\nu(y)}{2} \right)^l \sum_{j=0}^{m-2l} E_j(x, y). \end{aligned}$$

The double sum converges to $e^{-v(y)/2}E(x,y)$ as $m \to \infty$. Since V_κ is positive, $|E_j(x,y)| \le e^{\|x\|\cdot\|y\|}$, the terms in the double sum are dominated by

$$\sum_{l=0}^{\infty} \frac{\|y\|^{2l}}{l!2^l} \sum_{s=0}^{\infty} \frac{\|x\|^s\|y\|^s}{s!} = \exp\left(\frac{\|y\|^2}{2} + \|x\|\cdot\|y\|\right),$$

which is integrable with respect to $e^{-\|x\|^2/2}dx$. Hence, by the dominated convergence theorem,

$$p(y) = e^{v(y)/2} b_h \int_{\mathbb{R}^d} (e^{-\Delta_h/2} p(x)) E(x,y) h_\kappa(x)^2 e^{-\|x\|^2/2} dx.$$

Multiplying both sides by $e^{-v(y)}$ completes the proof of (6.1.2). Setting $p(x) = p_m(x) = \sum_{j=0}^{m} E_j(x,z)$ in (6.1.2), we see that (6.1.1) follows from $p_m(x) \to E(x,z)$ and $e^{-\Delta_h/2} p_m(x) \to e^{-v(z)/2}E(x,z)$, as $m \to \infty$, upon using the dominated convergence and Fubini theorems. □

The analog of the heat kernel for the Dunkl transform is defined by

$$q_t^\kappa(x) := (2t)^{-(\lambda_\kappa+1)} e^{-t\|x\|^2/4}, \qquad x \in \mathbb{R}^d, \quad t > 0, \tag{6.1.3}$$

where $\lambda_\kappa = \gamma_\kappa + \frac{d-2}{2}$ as before.

Corollary 6.1.4. *For $t > 0$,*

$$\mathscr{F}_\kappa q_t^\kappa(x) = e^{-t\|x\|^2}. \tag{6.1.4}$$

As an analog of the Fourier transform, we show that the Dunkl transform is an isometry of $L^2(\mathbb{R}^d, h_\kappa^2)$. First we give an orthogonal basis for $L^2(h_\kappa^2)$. Let $Y \in \mathscr{H}_n^d(h_\kappa^2)$. Define

$$\phi_m(Y;x) = L_m^{n+\lambda_\kappa}(\|x\|^2) Y(x) e^{-\|x\|^2/2}, \qquad x \in \mathbb{R}^d. \tag{6.1.5}$$

Proposition 6.1.5. *For $k,l,m,n \in \mathbb{N}_0$, the integral*

$$\frac{c_h}{(2\pi)^{d/2}} \int_{\mathbb{R}^d} \phi_m(Y_n;x)\phi_k(Y_l;x) h_\kappa^2(x) dx$$
$$= \delta_{mk}\delta_{nl} \frac{(\lambda_\kappa+1)_{n+m}}{2^{\lambda_\kappa+1} m} \frac{\omega_d}{\omega_d^\kappa} \int_{\mathbb{S}^{d-1}} [Y_n(x)]^2 h_\kappa^2(x) d\sigma.$$

Proof. Using spherical polar coordinates, the first integral equals

$$c_h \int_0^\infty L_m^{n+\lambda_\kappa}(r^2) L_k^{l+\lambda_\kappa}(r^2) e^{-r^2} r^{n+l+2\gamma_\kappa+d-1} dr \frac{2^{1-d/2}}{\Gamma(d/2)} \int_{\mathbb{S}^{d-1}} Y_n Y_l h_\kappa^2 d\sigma.$$

The second integral is zero if $n \ne l$. Assume $n = l$, make the change of variable $r^2 = t$. The first integral equals $(1/2)\delta_{mk}\Gamma(n+\lambda_k+1+m)/m!$, which gives the constant in the formula by (2.1.8). □

In particular, if $\{Y_{k,n}\}$ denotes an orthogonal basis of $\mathscr{H}_n^d(h_\kappa^2)$, then $\{\phi_m(Y_{k,n};x)\}$ forms an orthogonal basis of $L^2(\mathbb{R}^d, h_\kappa^2)$.

Theorem 6.1.6. *For* $m,n=0,1,2,\dots$, $Y\in\mathscr{H}_n^d(h_\kappa^2)$ *and* $y\in\mathbb{R}^d$,

$$(\mathscr{F}_\kappa\phi_m(Y))(y)=(-i)^{n+2m}\phi_m(Y;y).$$

Proof. By (6.1.2) and (3.1.17),

$$b_h\int_{\mathbb{R}^d}L_m^{n+\lambda_\kappa}(\|x\|^2/2)p(x)E(x,y)h_\kappa^2(x)e^{-\|x\|^2/2}\,dx$$
$$=(-1)^m(m!2^m)^{-1}e^{\nu(y)/2}\nu(y)^m p(y).$$

We change the argument in the Laguerre polynomial by using the identity

$$L_m^\alpha(t)=\sum_{j=0}^m 2^j\frac{(\alpha+1)_m}{(\alpha+1)_j}\frac{(-1)^{m-j}}{(m-j)!}L_j^\alpha(t/2),\qquad t\in\mathbb{R},$$

which can be derived from the generating function of the Laguerre polynomials. Using this expansion together with the above integral, we obtain

$$b_h\int_{\mathbb{R}^d}L_m^{n+\lambda_k}(\|x\|^2)p(x)e^{-\|x\|^2/2}E(x,y)h_\kappa^2(x)\,dx$$
$$=e^{\nu(y)/2}p(y)(-1)^m\frac{(n+\lambda_\kappa+1)_m}{m!}\sum_{j=0}^m\frac{(-m)_j}{(n+\lambda_\kappa+1)_j}\frac{(-\nu(y))^j}{j!}.$$

We now replace y by $-iy$ for $y\in\mathbb{R}^d$, so that $\nu(y)$ becomes $-\|y\|^2$ and $p(y)$ becomes $(-1)^m(-i)^n p(y)$, and the sum yields a Laguerre polynomial. Consequently, the integral equals $(-1)^m(-i)^n p(y)L_m^{n+\lambda_k}(\|y\|^2)e^{-\|y\|^2/2}$. □

Since the functions $\{\phi_m(Y):Y\in\mathscr{H}_n^d(h_k^2),m\ge 0\}$ constitutes a basis of $L^2(\mathbb{R}^d,h_\kappa^2)$ and the eigenvalues are powers of $i=\sqrt{-1}$, this proves the isometry properties, which are the analog of the Plancherel theorem stated below.

Theorem 6.1.7. *The Dunkl transform extends to an isometry of* $L^2(\mathbb{R}^d,h_\kappa^2)$ *onto itself. The square of the transform is the central involution; that is, if* $f\in L^2(\mathbb{R}^d,h_\kappa^2)$, $\mathscr{F}_\kappa f=g$, *then* $\mathscr{F}_\kappa g(x)=f(-x)$ *for almost all* $x\in\mathbb{R}^d$.

As a consequence, the inverse of the Dunkl transform is given by

$$f(x)=b_h\int_{\mathbb{R}^d}\mathscr{F}_\kappa f(y)E(ix,y)h_\kappa^2(y)dy,\tag{6.1.6}$$

which holds in $L^2(\mathbb{R}^d,h_\kappa^2)$. In particular, it holds for the Schwartz class of functions.

As another analog to the Fourier transform, the Dunkl transform of a radial function is also radial, and it can be expressed in terms of the Bessel function $J_\alpha(t)$, defined by

$$\begin{aligned}J_\alpha(t):=&\frac{(t/2)^\alpha}{\sqrt{\pi}\Gamma(\alpha+1/2)}\int_{-1}^1 e^{its}(1-s^2)^{\alpha-1/2}ds\\=&\left(\frac{t}{2}\right)^\alpha\sum_{n=0}^\infty\frac{(-1)^n}{n!\Gamma(n+\alpha+1)}\left(\frac{t}{2}\right)^{2n}.\end{aligned}\tag{6.1.7}$$

Theorem 6.1.8. *Suppose f is a radial function in $L^1(\mathbb{R}^d, h_\kappa^2)$; $f(x) = f_0(\|x\|)$ for almost all $x \in \mathbb{R}^d$. The Dunkl transform $\mathscr{F}_\kappa f$ is also radial and has the form $\mathscr{F}_\kappa f(x) = F_0(\|x\|)$ for all $x \in \mathbb{R}^d$ with*

$$F_0(\|x\|) = F_0(r) = \frac{1}{\omega_d r^{\lambda_\kappa}} \int_0^\infty f_0(s) J_{\lambda_\kappa}(rs) s^{\lambda_\kappa+1} ds.$$

Proof. Using polar coordinates and (3.2.6), we obtain

$$\begin{aligned}\mathscr{F}_\kappa f(y) &= b_h \int_0^\infty f_0(s) s^{d-1+2\gamma_\kappa} \int_{\mathbb{S}^{d-1}} E(sx', y) h_\kappa^2(x') d\sigma(x') ds \\ &= b_h c_{\lambda_\kappa} \omega_d^\kappa \int_0^\infty f_0(s) s^{2\lambda_\kappa+1} \int_{-1}^1 e^{is\|y\|t}(1-t^2)^{\lambda_\kappa - \frac{1}{2}} dt\, ds,\end{aligned}$$

from which the stated result follows by the definition of $J_\alpha(t)$ and, using (2.1.8), putting constants together. □

As a consequence, we see that the Dunkl transform of $f_0(\|x\|)$ is a Hankel transform in $\|x\|$. In general, the Hankel transform H_α is defined on the positive reals $\mathbb{R}_+$. For $\alpha > -1/2$,

$$H_\alpha f(s) := \frac{1}{\Gamma(\alpha+1)} \int_0^\infty f(r) \frac{J_\alpha(rs)}{(rs)^\alpha} r^{2\alpha+1} dr. \tag{6.1.8}$$

The inverse Hankel transform is given by

$$f(r) = \frac{1}{\Gamma(\alpha+1)} \int_0^\infty H_\alpha f(s) \frac{J_\alpha(rs)}{(rs)^\alpha} s^{2\alpha+1} ds, \tag{6.1.9}$$

which holds under mild conditions on f; for example, it holds if f is piecewise continuous and of bounded variation in every finite subinterval of $(0,\infty)$, and $\sqrt{r} f \in L^1(\mathbb{R}_+)$ ([65, p. 456]).

Example. Consider $h_\kappa(x) = |x|^\kappa$, $\kappa \ge 0$, on the real line. Then the group is $\mathbb{Z}_2$, and

$$E(x, -iy) = \Gamma(\kappa + 1/2)(|xy|/2)^{-\kappa+1/2} \left[J_{\kappa-1/2}(|xy|) - i \operatorname{sign}(xy) J_{\kappa+1/2}(|xy|)\right],$$

so that the Dunkl transform is related to the Hankel transform.

Indeed, in this case, the intertwining operator V_κ is given in (2.3.2), hence,

$$E(x, -iy) = c_\kappa \int_{-1}^1 e^{-isxy}(1+s)(1-s^2)^{\kappa-1} ds,$$

so that, integrating by parts,

$$\begin{aligned}E(x, -iy) &= c_\kappa \int_{-1}^1 e^{-isxy}(1-s^2)^{\kappa-1} ds - \frac{ix}{2\kappa} c_\kappa \int_{-1}^1 e^{-isxy}(1-s^2)^\kappa ds \\ &= \Gamma(\kappa+1/2)\left(|xy|/2\right)^{-\kappa+1/2}\left[J_{\kappa-1/2}(|x|) - i \operatorname{sign}(x) J_{\kappa+1/2}(|x|)\right].\end{aligned}$$

The following proposition is useful for dealing with the Dunkl transform of functions that involve h-harmonics.

Proposition 6.1.9. *Let* $f \in \mathscr{H}_n^d(h_\kappa^2)$, $n = 0, 1, 2, \ldots$. *If* $y \in \mathbb{R}^d$, $\rho > 0$, *then the function*

$$g(y) = \frac{\omega_d}{\omega_d^\kappa} \int_{\mathbb{S}^{d-1}} f(x) E(x, -iy\rho) h_\kappa^2(x) d\sigma(x)$$

satisfies $\Delta_h g = -\rho^2 g$ *and*

$$g(y) = (-i)^n \Gamma(\lambda_\kappa + 1) \Big(\frac{\rho \|y\|}{2}\Big)^{-\lambda_\kappa} f\Big(\frac{y}{\|y\|}\Big) J_{n+\lambda_\kappa}(\rho \|y\|).$$

Proof. First, let $y \in \mathbb{C}$. Since f is h-harmonic, $e^{-\Delta_h/2} f = f$. In the formula from (6.1.2),

$$b_h \int_{\mathbb{R}^d} f(x) E(x, y) h_\kappa^2(x) e^{-\|x\|^2/2} dx = e^{\nu(y)/2} f(y),$$

the part that is homogeneous of degree $n + 2m$ in y, $m = 0, 1, 2, \ldots$, yields the equation

$$b_h \int_{\mathbb{R}^d} f(x) E_{n+2m}(x, y) h_\kappa^2(x) e^{-\|x\|^2/2} dx = \frac{\nu(y)^m}{2^m m!} f(y).$$

Then, using the integral formula (2.1.7) and the fact that

$$\int_{\mathbb{S}^{d-1}} f(x) E_j(x, y) h_\kappa^2(x) d\sigma(x) = 0$$

if $j < n$ or $j \not\equiv n \bmod 2$, we conclude that

$$\begin{aligned} &\frac{\omega_d}{\omega_d^\kappa} \int_{\mathbb{S}^{d-1}} f(x) E(x, y) h_\kappa^2(x) d\sigma(x) \\ &\quad = \sum_{m=0}^{\infty} \frac{1}{2^{n+m}(d/2 + \gamma_\kappa)_{n+m}} c_h \int_{\mathbb{R}^d} f(x) E_{n+2m}(x, y) h_\kappa^2(x) e^{-\|x\|^2/2} dx. \end{aligned}$$

Replace y by $-i\rho y$ for $\rho > 0$, $y \in \mathbb{R}^d$. Let $A = n + \lambda_\kappa$. This leads to the expression for g in terms of J_A. To find $\Delta_h g$ we can interchange the integral and $\Delta_h^{(y)}$, because the resulting integral of a series $\sum_{n=0}^\infty \Delta_h^{(y)} E_n(x, -iy)$ converges absolutely. Indeed

$$\begin{aligned} \Delta_h^{(y)} E(x, -i\rho y) &= \Delta_h^{(y)} E(-i\rho x, y) \\ &= \sum_{j=1}^{N} (-i\rho x_j)^2 E(-i\rho x, y) = -\rho^2 \|x\|^2 E(x, -i\rho y). \end{aligned}$$

But $\|x\|^2 = 1$ on $\mathbb{S}^{d-1}$, and so $\Delta_h g = -\rho^2 g$. □

Let us denote by $\{Y_{j,n} : 1 \le j \le \dim \mathscr{H}_n^d(h_\kappa^2)\}$ an orthonormal basis of $\mathscr{H}_n^d(h_\kappa^2)$. We can prove a Paley-Wiener theorem for the Dunkl transform. Let $\mathscr{S}$ denote the space of Schwartz class of functions on $\mathbb{R}^d$.

Theorem 6.1.10. *Let $f \in \mathscr{S}$ and R be a positive number. Then f is supported in $\{x \in \mathbb{R}^d : \|x\| \le R\}$ if and only if for every j and n, the function*

$$F_{j,n}(\rho) = \rho^{-n} \int_{\mathbb{S}^{d-1}} \mathscr{F}_\kappa f(\rho x) Y_{j,n}(x) h_\kappa^2(x) d\sigma(x)$$

extends to an entire function of $\rho \in \mathbb{C}$ satisfying the estimate

$$|F_{j,n}(\rho)| \le c_{j,n} e^{R\|\operatorname{Im}\rho\|}.$$

Proof. By the definition of $\mathscr{F}_\kappa f$ and Proposition 6.1.9,

$$\begin{aligned}
&\int_{\mathbb{S}^{d-1}} \mathscr{F}_\kappa f(\rho x) Y_{j,n}(x) h_\kappa^2(x) d\sigma(x) \\
&= c \int_{\mathbb{R}^d} \Big(\int_{\mathbb{S}^{d-1}} E(y, -i\rho x) Y_{j,n}(x) h_\kappa^2(x) d\sigma(x) \Big) f(y) h_\kappa^2(y) dy \\
&= c \int_{\mathbb{R}^d} f(y) Y_{j,n}(y') \frac{J_{\lambda_k+n}(\rho\|y\|)}{(\rho\|y\|)^{\lambda_k}} h_\kappa^2(y) dy \\
&= c\rho^n \int_0^\infty f_{j,n}(r) \frac{J_{\lambda_k+n}(r\rho)}{(r\rho)^{\lambda_k+n}} r^{2\lambda_\kappa+2n+1} dr,
\end{aligned}$$

where c is a constant and

$$f_{j,n}(r) = \int_{\mathbb{S}^{d-1}} f(ry') Y_{j,n}(y') h_\kappa^2(y') d\sigma(y').$$

Thus, $F_{j,n}$ is the Hankel transform of order $\lambda_\kappa + n$ of the function $f_{j,n}(r)$. The theorem then follows from the Paley–Wiener theorem for the Hankel transform (see, for example, [36]). □

Corollary 6.1.11. *A function $f \in \mathscr{S}$ is supported in $\{x \in \mathbb{R}^d : \|x\| \le R\}$ if and only if $\mathscr{F}_\kappa f$ extends to an entire function of $\zeta \in \mathbb{C}^d$ which satisfies*

$$|\mathscr{F}_\kappa f(\zeta)| \le c e^{R\|\operatorname{Im}\zeta\|}.$$

Proof. The direct part follows from the fact that $E(x, -i\zeta)$ is entire and $|E(x, -i\zeta)| \le c e^{\|x\| \cdot \|\operatorname{Im}\zeta\|}$. For the converse we look at

$$\int_{\mathbb{S}^{d-1}} \mathscr{F}_\kappa f(\rho y') Y_{j,n}(y') h_\kappa^2(y') d\sigma(y'), \qquad \rho \in \mathbb{C}.$$

This is certainly entire and, from the proof of the previous theorem, has a zero of order n at the origin. Hence,

$$\rho^{-n} \int_{\mathbb{S}^{d-1}} \mathscr{F}_\kappa f(\rho y') Y_{j,n}(y') h_\kappa^2(y') d\sigma(y')$$

is an entire function of exponential type R, from which the converse follows from the theorem. □

6.2 Dunkl transform: L^1 theory

Let $\mathscr{S}$ denote the space of Schwartz functions on $\mathbb{R}^d$. The inversion formula of the Dunkl transform holds for $f \in \mathscr{S}$.

Theorem 6.2.1. *Let $f \in \mathscr{S}$. Then for $y \in \mathbb{R}^d$, $\mathscr{F}_\kappa \mathscr{D}_j f(y) = i y_j \mathscr{F}_\kappa f(y)$ for $j = 1, \ldots, d$. Furthermore, if $g_j(x) = x_j f(x)$, then $\mathscr{F}_\kappa g_j(y) = i \mathscr{D}_j \mathscr{F}_\kappa f(y)$, $y \in \mathbb{R}^d$. The operator $-i\mathscr{D}_j$ is densely defined on $L^2(\mathbb{R}^d, h_\kappa^2)$ and is self-adjoint.*

Proof. From the definition of $\mathscr{D}_j$, it is not difficult to prove that if $f, g \in \mathscr{S}$, then

$$\int_{\mathbb{R}^d} (\mathscr{D}_j f) g h_\kappa^2 dx = -\int_{\mathbb{R}^d} f(\mathscr{D}_j g) h_\kappa^2 dx, \quad j = 1, \ldots, d.$$

For fixed $y \in \mathbb{R}^d$, put $g(x) = E(x, -iy)$ in the above identity. Then $\mathscr{D}_j g(x) = -i y_j E(x, -iy)$ and $\mathscr{F}_\kappa \mathscr{D}_j f(y) = (-1)(-i y_j) \mathscr{F}_\kappa f(y)$. The multiplication operator defined by $M_j f(y) = y_j f(y)$, $j = 1, \ldots, d$, is densely defined and self-adjoint on $L^2(\mathbb{R}^d, h_\kappa^2)$. Furthermore, $-i\mathscr{D}_j$ is the inverse image of M_j under the Dunkl transform. □

In particular, this shows that if $f \in \mathscr{S}$, then $\mathscr{F}_\kappa f \in \mathscr{S}$. The assumption $f \in \mathscr{S}$ can of course be substantially relaxed.

Theorem 6.2.2. *If $f \in L^1(\mathbb{R}^d; h_\kappa^2)$, then $f \in C_0(\mathbb{R}^d)$.*

Proof. The space $\mathscr{S}$ is dense in $L^1(\mathbb{R}^d; h_\kappa^2)$. For each $f \in L^1(\mathbb{R}^d; h_\kappa^2)$, there are functions $f_n \in \mathscr{S}$ such that $\|f - f_n\|_{\kappa,1} \to 0$. Since $\mathscr{F}_\kappa f_n \in \mathscr{S} \subset C_0(\mathbb{R}^d)$ and $\mathscr{F}_\kappa f_n$ converges uniformly to $\mathscr{F}_\kappa f$ by $\|\mathscr{F}_\kappa f\|_\infty \le \|f\|_{\kappa,1}$, it follows that $\mathscr{F}_\kappa f \in C_0(\mathbb{R}^d)$. □

We want to show that if both f and $\mathscr{F}_\kappa f$ are in $L^1(\mathscr{R}^d; h_\kappa^2)$, then the inversion theorem holds. For this purpose, we use a generalized convolution operator. Recall that the usual convolution $f * g$ is defined in terms of the translation $\tau_y f = f(\cdot - y)$ of $\mathbb{R}^d$, which satisfies, taking the Fourier transform, $\mathscr{F}_\kappa \tau_y f(x) = e^{-i\langle x,y\rangle} \mathscr{F}_\kappa f(x)$. The translation operator works for the Lebesgue measure since it leaves $L^1(\mathbb{R}^d)$ invariant. It is not obvious what operation is translation invariant for $L^1(\mathbb{R}^d, h_\kappa^2)$. We define it in the Dunkl transform side.

Definition 6.2.3. Let $y \in \mathbb{R}^d$ be given. The generalized translation operator $f \mapsto \tau_y f$ is defined on $L^2(\mathbb{R}^d; h_\kappa^2)$ by the relation

$$\mathscr{F}_\kappa \tau_y f(x) := E(y, -ix) \mathscr{F}_\kappa f(x), \qquad x \in \mathbb{R}^d. \tag{6.2.1}$$

The definition makes sense as the Dunkl transform is an isometry of $L^2(\mathbb{R}^d; h_\kappa^2)$ onto itself and the function $E(y, -ix)$ is bounded. However, none of the other properties of the usual translation operator is obvious; for example, boundedness in L^p, translation invariance, or positivity. Some of these properties will be studied below and in the next section.

We start with an example. For $f \in \mathscr{S}$ we can write

$$\tau_y f(x) = b_h \int_{\mathbb{R}^d} E(ix, \xi) E(-iy, \xi) \mathscr{F}_\kappa f(\xi) h_\kappa^2(\xi) d\xi. \tag{6.2.2}$$

Proposition 6.2.4. *For $t > 0$ and $x \in \mathbb{R}^d$,*

$$\tau_y e^{-t\|x\|^2} = e^{-t(\|x\|^2+\|y\|^2)} E(\sqrt{2t}x, \sqrt{2t}y). \tag{6.2.3}$$

Proof. This follows immediately from (6.1.1), (6.1.4) and (6.2.2). □

Proposition 6.2.5. *Assume that $f, g \in \mathscr{S}$. Then*

(1) $\displaystyle\int_{\mathbb{R}^d} \tau_y f(\xi) g(\xi) h_\kappa^2(\xi) d\xi = \int_{\mathbb{R}^d} f(\xi) \tau_{-y} g(\xi) h_\kappa^2(\xi) d\xi.$

(2) $\tau_y f(x) = \tau_{-x} f(-y).$

Proof. The property (2) follows from the definition, since $E(\lambda x, \xi) = E(x, \lambda \xi)$ for any $\lambda \in \mathbb{C}$. If $f, g \in \mathscr{S}$, then both integrals in (1) are well defined. From the definition,

$$\begin{aligned}\int_{\mathbb{R}^d} \tau_y f(\xi) g(\xi) h_\kappa^2(\xi) d\xi &= b_h \int_{\mathbb{R}^d} \left(\int_{\mathbb{R}^d} E(ix, \xi) E(-iy, \xi) \mathscr{F}_\kappa f(\xi) h_\kappa^2(\xi) d\xi \right) g(x) h_\kappa^2(x) dx \\ &= \int_{\mathbb{R}^d} \mathscr{F}_\kappa f(\xi) \mathscr{F}_\kappa g(-\xi) E(-iy, \xi) h_\kappa^2(\xi) d\xi\end{aligned}$$

by the inversion theorem applied to g. We also have

$$\begin{aligned}\int_{\mathbb{R}^d} f(\xi) \tau_{-y} g(\xi) h_\kappa^2(\xi) d\xi &= b_h \int_{\mathbb{R}^d} \left(\int_{\mathbb{R}^d} E(ix, \xi) E(iy, \xi) \mathscr{F}_\kappa g(\xi) h_\kappa^2(\xi) d\xi \right) f(x) h_\kappa^2(x) dx \\ &= \int_{\mathbb{R}^d} \mathscr{F}_\kappa f(-\xi) \mathscr{F}_\kappa g(\xi) E(iy, \xi) h_\kappa^2(\xi) d\xi \\ &= \int_{\mathbb{R}^d} \mathscr{F}_\kappa f(\xi) \mathscr{F}_\kappa g(-\xi) E(-iy, \xi) h_\kappa^2(\xi) d\xi\end{aligned}$$

by the inversion theorem applied to f. This proves (1). □

Proposition 6.2.6. *Let $f \in \mathscr{S}$ be supported in $\{x \in \mathbb{R}^d : \|x\| \le R\}$. Then $\tau_y f$ is supported in $\{x \in \mathbb{R}^d : \|x\| \le R + \|y\|\}$.*

Proof. Let $g(x) = \tau_y f(x)$. Then, by Corollary 6.1.11, $\mathscr{F}_\kappa g(\xi) = E(y, -i\xi) \mathscr{F}_\kappa f(\xi)$ extends to $\mathbb{C}^d$ as an entire function of type $R + \|y\|$. Hence, the stated result follows from Corollary 6.1.11. □

Theorem 6.2.7. *If $f \in C_0^\infty(\mathbb{R}^d)$ is supported in $\|x\| \le R$, then*

$$\|\tau_y f - f\|_p \le c_f \|y\| (R + \|y\|)^{\frac{2\lambda_\kappa + 2}{p}}$$

for $1 \le p \le \infty$. Consequently, $\lim_{y \to 0} \|\tau_y f - f\|_{\kappa,p} = 0$.

Proof. From the definition we have

$$\tau_y f(x) - f(x) = b_h \int_{\mathbb{R}^d} (E(y, -i\xi) - 1) E(x, i\xi) \mathscr{F}_\kappa f(\xi) h_\kappa^2(\xi) d\xi.$$

Using the mean value theorem and estimates on the derivatives of $E(x,i\xi)$, we obtain the estimate

$$\|\tau_y f - f\|_\infty \le c\|y\| \int_{\mathbb{R}^d} \|\xi\| \, |\mathscr{F}_\kappa f(\xi)| h_\kappa^2(\xi) d\xi.$$

As f is supported in $\|x\| \le R$ and $\tau_y f$ is supported in $\|x\| \le (R+\|y\|)$, we can restrict the integration domain above to $\|x\| \le (R+\|y\|)$ and conclude, accordingly, that

$$\|\tau_y f - f\|_p \le c_f \|y\| (R+\|y\|)^{\frac{2\lambda_\kappa+2}{p}},$$

which goes to zero as y goes to zero. □

The generalized translation operator is used to define a convolution structure:

Definition 6.2.8. For $f,g \in L^2(\mathbb{R}^d;h_\kappa^2)$ we define

$$f *_\kappa g(x) := b_h \int_{\mathbb{R}^d} f(y) \tau_x \widetilde{g}(y) h_\kappa^2(y) dy,$$

where $\widetilde{g}(y) := g(-y)$.

Since $\tau_x \widetilde{g} \in L^2(\mathbb{R}^d;h_\kappa^2)$ the above convolution is well defined. By (6.2.1), we can also write the definition as

$$f *_\kappa g(x) = b_h \int_{\mathbb{R}^d} \mathscr{F}_\kappa f(\xi) \mathscr{F}_\kappa g(\xi) E(ix,\xi) h_\kappa^2(\xi) d\xi. \tag{6.2.4}$$

Recall that $q_t^\kappa(x) = (2t)^{-\lambda_k+1} e^{-t\|x\|^2}$ denotes the heat kernel. Changing variable shows that $b_h \int_{\mathbb{R}^d} q_t^\kappa(x) h_\kappa^2(x) dx = 1$.

Lemma 6.2.9. *For $f \in L^1(\mathbb{R}^d;h_\kappa^2)$,*

$$\lim_{t\to 0^+} \|f *_\kappa q_t^\kappa - f\|_{\kappa,1} = 0.$$

Proof. By (1) in Proposition 6.2.5 with $f = q_t^\kappa$ and $g = 1$, $b_h \int_{\mathbb{R}^d} \tau_x q_t^\kappa(y) h_\kappa^2(y) dy = 1$. Since $\tau_u q_t^\kappa \ge 0$ by (6.2.3), it follows then that

$$\|f * q_t^\kappa\|_{\kappa,1} \le \|f\|_{\kappa,1}.$$

For a given $\varepsilon > 0$ we choose $g \in \mathscr{S}$ such that $\|g-f\|_{\kappa,1} < \varepsilon/3$. The triangle inequality then leads to

$$\|f *_\kappa q_t^\kappa - f\|_{\kappa,1} \le \frac{2}{3}\varepsilon + \|g *_\kappa q_t^\kappa - g\|_{\kappa,1}.$$

Since $g \in \mathscr{S}$, it follows that

$$g *_\kappa q_t^\kappa(x) = \int_{\mathbb{R}^d} g(y) \tau_x(\widetilde{q_t^\kappa})(y) h_\kappa^2(y) dy = \int_{\mathbb{R}^d} \tau_{-x} g(y) q_t^\kappa(-y) h_\kappa^2(y) dy.$$

We also know that $\tau_{-x} g(y) = \tau_{-y} g(x)$. Therefore,

$$g *_\kappa q_t^\kappa(x) = \int_{\mathbb{R}^d} \tau_y g(x) q_t^\kappa(y) h_\kappa^2(y) dy.$$

In view of this,

$$g *_\kappa q_t^\kappa(x) - g(x) = \int_{\mathbb{R}^d} (\tau_y g(x) - g(x))\, q_t^\kappa(y) h_\kappa^2(y) dy,$$

which implies then

$$\|g *_\kappa q_t^\kappa - g\|_{\kappa,1} \le \int_{\mathbb{R}^d} \|\tau_y g - g\|_{\kappa,1} |q_t^\kappa(y)| h_\kappa^2(y) dy.$$

If g is supported in $\|x\| \le R$, then the estimate in Theorem 6.2.7 gives

$$\begin{aligned} \|g *_\kappa q_t^\kappa - g\|_{\kappa,1} &\le c \int_{\mathbb{R}^d} \|y\| \, (R + \|y\|)^{2\lambda_\kappa + 2} \, |q_t^\kappa(y)| h_\kappa^2(y) dy \\ &\le ct \int_{\mathbb{R}^d} \|y\| \, (R + \|ty\|)^{2\lambda_k + 1} \, e^{-\|y\|^2} h_\kappa^2(y) dy, \end{aligned}$$

which can be made smaller than $\varepsilon/3$ by choosing ε small. This completes the proof of the lemma. □

Theorem 6.2.10. *If both f and $\mathscr{F}_\kappa f \in L^1(\mathbb{R}^d, h_\kappa^2)$, then for almost all $x \in \mathbb{R}^d$,*

$$f(x) = b_h \int_{\mathbb{R}^d} \mathscr{F}_\kappa f(y) E(ix, y) h_\kappa^2(y) dy.$$

Proof. Since $\mathscr{F}_\kappa q_t^\kappa(x) = e^{-\|x\|^2}$, for $f \in \mathscr{S}$ we have

$$f * q_t^\kappa(x) = b_h \int_{\mathbb{R}^d} \mathscr{F}_\kappa f(y) e^{-t\|y\|^2} E(ix, y) h_\kappa^2(y) dy.$$

This extends to $f \in L^1(h_\kappa^2; \mathbb{R}^d)$ since the convolution operator extends to $L^1(h_\kappa^2; \mathbb{R}^d)$ as a bounded operator by $\|f *_\kappa q_t^\kappa\|_{\kappa,1} \le \|f\|_{\kappa,1}$. Letting $t \to 0^+$, applying Lemma 6.2.9 to the left-hand side and the dominant convergence theorem to the right-hand side, we see that the inversion formula follows almost everywhere. □

For the convenience of applications, we summarize some of the most useful properties of the Dunkl transform established in this and the preceding section as follows.

Theorem 6.2.11. (i) *If $f \in L^1(\mathbb{R}^d; h_\kappa^2)$, then $\mathscr{F}_\kappa f \in C(\mathbb{R}^d)$ and $\lim_{\|\xi\| \to \infty} \mathscr{F}_\kappa f(\xi) = 0$.*

(ii) *The Dunkl transform $\mathscr{F}_\kappa$ is an isomorphism of the Schwartz class $\mathscr{S}(\mathbb{R}^d)$ onto itself, and $\mathscr{F}_\kappa^2 f(x) = f(-x)$.*

(iii) *The Dunkl transform $\mathscr{F}_\kappa$ on $\mathscr{S}(\mathbb{R}^d)$ extends uniquely to an isometric isomorphism on $L^2(\mathbb{R}^d; h_\kappa^2)$, i.e., $\|f\|_{\kappa,2} = \|\mathscr{F}_\kappa f\|_{\kappa,2}$.*

(iv) *If f and $\mathscr{F}_\kappa f$ are both in $L^1(\mathbb{R}^d; h_\kappa^2)$, then the following inverse formula holds:*

$$f(x) = c_\kappa \int_{\mathbb{R}^d} \mathscr{F}_\kappa f(y) E_\kappa(ix, y) h_\kappa^2(y) \, dy, \quad x \in \mathbb{R}^d.$$

(v) *If* $f, g \in L^2(\mathbb{R}^d; h_\kappa^2)$, *then*

$$\int_{\mathbb{R}^d} \mathscr{F}_\kappa f(x) g(x)\, h_\kappa^2(x)\, dx = \int_{\mathbb{R}^d} f(x) \mathscr{F}_\kappa g(x)\, h_\kappa^2(x)\, dx.$$

(vi) *Given* $\varepsilon > 0$, *let* $f_\varepsilon(x) = \varepsilon^{-2-2\gamma_\kappa} f(\varepsilon^{-1}x)$. *Then* $\mathscr{F}_\kappa f_\varepsilon(\xi) = \mathscr{F}_\kappa f(\varepsilon\xi)$.

(vii) *If* $f(x) = f_0(\|x\|)$ *is radial, then* $\mathscr{F}_\kappa f(\xi) = H_{\lambda_\kappa} f_0(\|\xi\|)$ *is again a radial function, where* H_α *denotes the Hankel transform defined by*

$$H_\alpha g(s) = \frac{1}{\Gamma(\alpha+1)} \int_0^\infty g(r) \frac{J_\alpha(rs)}{(rs)^\alpha} r^{2\alpha+1}\, dr,$$

and J_α *denotes the Bessel function of the first kind.*

6.3 Generalized translation operator

Here we study properties of the generalized translation operator. It is convenient to define a class of functions on which (6.2.1) holds pointwisely:

$$A_\kappa(\mathbb{R}^d) := \{f \in L^1(\mathbb{R}^d; h_\kappa^2) : \mathscr{F}_\kappa f \in L^1(\mathbb{R}^d; h_\kappa^2)\}.$$

This is a subspace of the intersection of $L^1(\mathbb{R}^d; h_\kappa^2)$ and L^∞ and, hence, a subspace of $L^2(\mathbb{R}^d; h_\kappa^2)$. The assumption on f in Proposition 6.2.5 can be relaxed as follows:

Proposition 6.3.1. *Assume that* $f \in A_\kappa(\mathbb{R}^d)$ *and* $g \in L^1(\mathbb{R}^d; h_\kappa^2)$ *is bounded. Then*

(1) $\displaystyle\int_{\mathbb{R}^d} \tau_y f(\xi) g(\xi) h_\kappa^2(\xi) d\xi = \int_{\mathbb{R}^d} f(\xi) \tau_{-y} g(\xi) h_\kappa^2(\xi) d\xi.$

(2) $\tau_y f(x) = \tau_{-x} f(-y)$.

Proof. The proof of (2) is the same as before. If both f and g are in $A_\kappa(\mathbb{R}^d)$, then the proof of Proposition 6.2.5 works. Suppose now $f \in A_\kappa(\mathbb{R}^d)$, $g \in L^1(\mathbb{R}^d; h_\kappa^2) \cap L^\infty$. Since $g \in L^2(\mathbb{R}^d; h_\kappa^2)$, $\tau_y g$ is defined as an L^2 function. As f is in $L^2(\mathbb{R}^d; h_\kappa^2)$ and bounded, both integrals are finite. The relation

$$\int_{\mathbb{R}^d} f(\xi) \mathscr{F}_\kappa g(\xi) h_\kappa^2(\xi) d\xi = \int_{\mathbb{R}^d} \mathscr{F}_\kappa f(\xi) g(\xi) h_\kappa^2(\xi) d\xi,$$

which is true for Schwartz class functions, remains true for $f, g \in L^2(\mathbb{R}^d; h_\kappa^2)$ as well. Using this we get

$$\begin{aligned} \int_{\mathbb{R}^d} \tau_y f(x) g(x) h_\kappa^2(x) dx &= \int_{\mathbb{R}^d} \tau_y f(-x) g(-x) h_\kappa^2(x) dx \\ &= \int_{\mathbb{R}^d} E(y, -i\xi) \mathscr{F}_\kappa f(\xi) \mathscr{F}_\kappa g(-\xi) h_\kappa^2(\xi) d\xi. \end{aligned}$$

By the same argument, the integral on the right-hand side is also given by the same expression. Hence (1) is proved. □

6.3.1 Translation operator on radial functions

A function is called radial, if it depends only on $\|x\|$. For such a function, there is an explicit expression of the generalized translation operator.

Theorem 6.3.2. *Let $f(x) = f_0(\|x\|)$. Assume $f \in A_\kappa(\mathbb{R}^d)$. Then for almost every $x \in \mathbb{R}^d$,*

$$\tau_y f(x) = V_\kappa \left[f_0 \left(\sqrt{\|x\|^2 + \|y\|^2 - 2\|x\| \, \|y\| \langle x', \cdot \rangle} \right) \right] \left(\frac{y}{\|y\|} \right) \text{ where } x' = \frac{x}{\|x\|}.$$

Proof. Since f is radial, its Dunkl transform is also a radial function, which we denote by $\mathscr{F}_\kappa f_0(r)$. Using (6.2.1), the property (iv) of Theorem 6.2.11, and the spherical-polar coordinates $\xi = r\xi'$, we get

$$\tau_y f(x) = b_h \int_0^\infty r^{2\lambda_\kappa + 1} \left[\int_{\mathbb{S}^{d-1}} E(x, -ir\xi') E(y, ir\xi') h_\kappa^2(\xi') d\sigma(\xi') \right] \mathscr{F}_\kappa f_0(r) dr.$$

We compute the inner integral first. For each $y' \in \mathbb{S}^{d-1}$, the reproducing kernel $P_n(h_\kappa^2; y', \cdot)$ is an element of $\mathscr{H}_n^d(h_\kappa^2)$. Hence, Proposition 6.1.9 shows that

$$\begin{aligned} &\frac{\omega_d}{\omega_d^\kappa} \int_{\mathbb{S}^{d-1}} E(x, -ir\xi') P_n(h_\kappa^2; y', \xi') h_\kappa^2(\xi') d\sigma(\xi') \\ &\qquad = (-i)^n 2^{\lambda_\kappa} (r\|x\|)^{-\lambda_\kappa} J_{n+\lambda_\kappa}(r\|x\|) P_n(h_\kappa^2; y', x'), \end{aligned}$$

where $x' = x/\|x\|$, which implies that in $L^2(\mathbb{S}^{d-1}, h_\kappa^2)$,

$$E(x, -ir\xi') = c_{\kappa,d} \sum_{n=0}^\infty (-i)^n (r\|x\|)^{-\lambda_\kappa} J_{n+\lambda_\kappa}(r\|x\|) P_n(h_\kappa^2; \xi', x').$$

Replacing ξ' by $-\xi'$ gives the expansion of $E(x, ir\xi')$. Hence, using the reproducing property of $P_n(h_\kappa^2; \cdot, \cdot)$, we get

$$\begin{aligned} &\frac{\omega_d}{\omega_d^\kappa} \int_{\mathbb{S}^{d-1}} E(x, -ir\xi') E(y, ir\xi') h_\kappa^2(\xi') d\sigma(\xi') \\ &\quad = c'_{\kappa,d} \sum_{n=0}^\infty \frac{J_{n+\lambda_\kappa}(r\|x\|)}{(r\|x\|)^{\lambda_\kappa}} \frac{J_{n+\lambda_\kappa}(r\|y\|)}{(r\|y\|)^{\lambda_\kappa}} P_n(h_\kappa^2; y', x') \\ &\quad = c'_{\kappa,d} V_\kappa \left[\sum_{n=0}^\infty \frac{J_{n+\lambda_\kappa}(r\|x\|)}{(r\|x\|)^{\lambda_\kappa}} \frac{J_{n+\lambda_\kappa}(r\|y\|)}{(r\|y\|)^{\lambda_\kappa}} \frac{n+\lambda_\kappa}{\lambda_\kappa} C_n^{\lambda_\kappa}(\langle x', \cdot \rangle) \right] (y'). \end{aligned}$$

By the addition formula for Bessel functions ([2, p. 215]), the last expression is equal to

$$c V_\kappa \left[\frac{J_{\lambda_\kappa}(r\sqrt{\|x\|^2 + \|y\|^2 - 2\|x\| \, \|y\| \langle x', \cdot \rangle})}{r^{\lambda_\kappa} (\|x\|^2 + \|y\|^2 - 2\|x\| \, \|y\| \langle x', \cdot \rangle)^{\lambda_\kappa/2}} \right] (y'),$$

where c is a constant. Consequently, we conclude that

$$\begin{aligned}\tau_y f(x) &= cV_\kappa\left[\int_0^\infty r^{2\lambda_\kappa+1}\frac{J_{\lambda_\kappa}(rz(x,y,\cdot))}{(rz(x,y,\cdot))^{\lambda_\kappa}}\mathscr{F}_\kappa f_0(r)dr\right](y')\\ &= cV_\kappa\left[H_{\lambda_\kappa}\mathscr{F}_\kappa f_0(z(x,y,\cdot))\right](y'),\end{aligned}$$

where $z(x,y,\cdot) = \sqrt{\|x\|^2+\|y\|^2-2\|x\|\,\|y\|\langle x',\cdot\rangle}$ and c is a constant independent of f. By Theorem 6.1.8, $\mathscr{F}_\kappa f(x) = cH_{\lambda_\kappa} f_0(\|x\|)$, thus it follows from the inversion formula of the Hankel transform (6.1.9) that $\tau_y f(x) = cV_\kappa[f_0(z(x,y,\cdot))](y')$. The constant c can be determined by setting $f(x)\equiv 1$. This completes the proof. □

The condition in Theorem 6.3.2 can be relaxed somewhat, see Lemma 7.2.4 below.
An immediate consequence of the explicit expression of τ_y is the following:

Theorem 6.3.3. *Let $f\in A_\kappa(\mathbb{R}^d)$ be radial and nonnegative. Then $\tau_y f\geq 0$, $\tau_y f\in L^1(\mathbb{R}^d;h_\kappa^2)$ and*

$$\int_{\mathbb{R}^d}\tau_y f(x)h_\kappa^2(x)dx = \int_{\mathbb{R}^d} f(x)h_\kappa^2(x)dx. \tag{6.3.1}$$

Proof. As f is radial, the explicit formula in Proposition 6.3.2 shows that $\tau_y f\geq 0$ since V_κ is a positive operator. Taking $g(x) = e^{-t\|x\|^2}$ and making use of (6.2.3) we obtain from (1) of Proposition 6.2.5 that

$$\int_{\mathbb{R}^d}\tau_y f(x)e^{-t\|x\|^2}h_\kappa^2(x)dx = \int_{\mathbb{R}^d} f(x)e^{-t(\|x\|^2+\|y\|^2)}E(\sqrt{2t}x,\sqrt{2t}y)h_\kappa^2(x)dx.$$

As $|E(x,y)|\leq e^{\|x\|\,\|y\|}$, we can take limit as $t\to 0$ to get

$$\lim_{t\to 0}\int_{\mathbb{R}^d}\tau_y f(x)e^{-t\|x\|^2}h_\kappa^2(x)dx = \int_{\mathbb{R}^d} f(x)h_\kappa^2(x)dx.$$

Since $\tau_y f\geq 0$, the monotone convergence theorem applied to the integral on the left completes the proof. □

We can relax the conditions on f as follows.

Theorem 6.3.4. *Let $f\in L^1(\mathbb{R}^d;h_\kappa^2)\cap L^\infty$ be radial and nonnegative. Then $\tau_y f\geq 0$, $\tau_y f\in L^1(\mathbb{R}^d;h_\kappa^2)$ and* (6.3.1) *holds.*

Proof. Since f is radial and nonnegative, the convolution $f *_\kappa q_t^\kappa$ is also radial and nonnegative. Since f is both in $L^1(\mathbb{R}^d;h_\kappa^2)$ and $L^2(\mathbb{R}^d;h_\kappa^2)$, $f *_\kappa q_t^\kappa\in L^1(\mathbb{R}^d,h_\kappa^2)$ because $q_t^\kappa\in A_\kappa(\mathbb{R}^d)$ and, by the Plancherel theorem and Hölder's inequality, $\|\mathscr{F}_\kappa f *_\kappa q_t^\kappa\|_{\kappa,1} = \|\mathscr{F}_\kappa f\cdot\mathscr{F}_\kappa q_t^\kappa\|_{\kappa,1}\leq\|f\|_{\kappa,2}\|q_t^\kappa\|_{\kappa,2}$. Hence, $f *_\kappa q_t^\kappa\in A_\kappa(\mathbb{R}^d)$. Thus, by Theorem 6.3.3, $\tau_y(f *_\kappa q_t^\kappa)(x)\geq 0$. Since $f\in L^2(\mathbb{R}^d;h_\kappa^2)$, it is easy to see that $\|f *_\kappa q_t^\kappa - f\|_{\kappa,2}\to 0$. Since τ_y is bounded on $L^2(\mathbb{R}^d;h_\kappa^2)$, we have $\tau_y(f *_\kappa q_t)\to\tau_y f$ in $L^2(\mathbb{R}^d;h_\kappa^2)$ as $t\to 0$. By passing to a subsequence if necessary, we can assume that the convergence is also almost everywhere. This gives us

$$\lim_{t\to 0}\tau_y(f *_\kappa q_t)(x) = \tau_y g(x)\geq 0$$

for almost every x. Since $\tau_y f$ is nonnegative, we let $t \to 0$ and can apply the monotone convergence theorem to

$$\int_{\mathbb{R}^d} \tau_y f(x) e^{-t\|x\|^2} h_\kappa^2(x)dx = \int_{\mathbb{R}^d} f(x) e^{-t(\|x\|^2+\|y\|^2)} E(\sqrt{2t}x, \sqrt{2t}y) h_\kappa^2(x)dx$$

and get (6.3.1). □

Let $L^p_{\text{rad}}(\mathbb{R}^d; h_\kappa^2)$ denote the space of all radial functions in $L^p(\mathbb{R}^d; h_\kappa^2)$.

Theorem 6.3.5. *The generalized translation operator τ_y can be extended to all radial functions in $L^p(\mathbb{R}^d; h_\kappa^2)$, $1 \le p \le 2$, and $\tau_y : L^p_{\text{rad}}(\mathbb{R}^d, h_\kappa^2) \to L^p(\mathbb{R}^d; h_\kappa^2)$ is a bounded operator.*

Proof. For a radial function $f \in L^1(\mathbb{R}^d; h_\kappa^2) \cap L^\infty$, the inequality $-|f| \le f \le |f|$ together with the nonnegativity of τ_y on radial functions in $L^1(\mathbb{R}^d; h_\kappa^2) \cap L^\infty$ shows that $|\tau_y f(x)| \le \tau_y |f|(x)$. Hence

$$\int_{\mathbb{R}^d} |\tau_y f(x)| h_\kappa^2(x)dx \le \int_{\mathbb{R}^d} |f|(x) h_\kappa^2(x)dx \le \|f\|_{\kappa,1}.$$

We also have $\|\tau_y f\|_{\kappa,2} \le \|f\|_{\kappa,2}$. By interpolation between L^1 and L^2, then $\|\tau_y f\|_{\kappa,p} \le \|f\|_{\kappa,p}$ for all $1 \le p \le 2$ and all $f \in L^p_{\text{rad}}(\mathbb{R}^d; h_\kappa^2)$. This proves the theorem. □

Theorem 6.3.6. *For every $f \in L^1_{\text{rad}}(\mathbb{R}^d; h_\kappa^2)$,*

$$\int_{\mathbb{R}^d} \tau_y f(x) h_\kappa^2(x)dx = \int_{\mathbb{R}^d} f(x) h_\kappa^2(x)dx.$$

Proof. Choose radial functions $f_n \in A_\kappa(\mathbb{R}^d)$ such that $f_n \to f$ and $\tau_y f_n \to \tau_y f$ in $L^1(\mathbb{R}^d; h_\kappa^2)$. Since

$$\int_{\mathbb{R}^d} \tau_y f_n(x) g(x) h_\kappa^2(x)dx = \int_{\mathbb{R}^d} f_n(x) \tau_{-y} g(x) h_\kappa^2(x)dx$$

for every $g \in A_\kappa(\mathbb{R}^d)$ we get, taking limit as n tends to infinity,

$$\int_{\mathbb{R}^d} \tau_y f(x) g(x) h_\kappa^2(x)dx = \int_{\mathbb{R}^d} f(x) \tau_{-y} g(x) h_\kappa^2(x)dx.$$

Now take $g(x) = e^{-t\|x\|^2}$ and take the limit as t goes to 0. Since $\tau_y f \in L^1(\mathbb{R}^d; h_\kappa^2)$, the dominated convergence theorem shows that

$$\int_{\mathbb{R}^d} \tau_y f(x) h_\kappa^2(x)dx = \int_{\mathbb{R}^d} f(x) h_\kappa^2(x)dx$$

for $f \in L^1(\mathbb{R}^d; h_\kappa^2)$. □

For non-radial functions, say in the Schwartz class, it is known that τ_y is not positive when the group G is either $\mathbb{Z}_2^d$ or the symmetric group, and this should be the case for all other reflection groups. It remains an open problem if $\tau_y f$ can be defined for all $f \in L^1(\mathbb{R}^d; h_\kappa^2)$ when the group G is not $\mathbb{Z}_2^d$. The case $G = \mathbb{Z}_2^d$ is discussed in the next subsection.

6.3.2 Translation operator for $G = \mathbb{Z}_2^d$

Recall that the weight function h_κ, invariant under the group $\mathbb{Z}_2^d$, takes the form

$$h_\kappa(x) = \prod_{i=1}^{d} |x_i|^{\kappa_i}, \qquad \kappa_i \geq 0.$$

The explicit formula (2.3.2) for the intertwining operator V_κ for $\mathbb{Z}_2^d$ allows us to derive an explicit formula for τ_y. Let us first consider the case $d = 1$.

Theorem 6.3.7. *For $G = \mathbb{Z}_2$ and $h_\kappa(x) = |x|^\kappa$ on $\mathbb{R}$,*

$$\begin{aligned}\tau_y f(x) =& \frac{1}{2}\int_{-1}^{1} f\left(\sqrt{x^2+y^2-2xyt}\right)\left(1+\frac{x-y}{\sqrt{x^2+y^2-2xyt}}\right)\Phi_\kappa(t)dt \\ &+\frac{1}{2}\int_{-1}^{1} f\left(-\sqrt{x^2+y^2-2xyt}\right)\left(1-\frac{x-y}{\sqrt{x^2+y^2-2xyt}}\right)\Phi_\kappa(t)dt, \qquad (6.3.2)\end{aligned}$$

where $\Phi_\kappa(t) = b_\kappa(1+t)(1-t^2)^{\kappa-1}$.

Proof. In this case, f radial means that f is an even function. Using the explicit formula of V_κ in (2.3.2), the formula in Theorem 6.3.2 shows that if f is even, then

$$\tau_y f(x) = \int_{-1}^{1} f\left(\sqrt{x^2+y^2-2xyt}\right)\Phi_\kappa(t)dt.$$

Making use of the fact that the derivative of an even function is odd, we derive a formula for $\tau_y f'$ using the fact that $\mathscr{D}\tau_y = \tau_y\mathscr{D}$. In this simple case the Dunkl operator $\mathscr{D}$ is given by

$$\mathscr{D}f(x) = f'(x) + \kappa\frac{f(x)-f(-x)}{x}.$$

On the one hand, since a radial function is invariant under the difference part, we have

$$\mathscr{D}\tau_y f(x) = \tau_y\mathscr{D}f(x) = \tau_y f'(x).$$

On the other hand, for f even, a simple computation shows that

$$\begin{aligned}\frac{\tau_y f(x) - \tau_y f(-x)}{x} &= \frac{1}{x}\int_{-1}^{1}\left(\int_{-t}^{t}\frac{d}{ds}f\left(\sqrt{x^2+y^2-2xys}\right)ds\right)\Phi_\kappa(t)dt \\ &= -y\int_{-1}^{1}\left(\int_{-t}^{t}\frac{f'\left(\sqrt{x^2+y^2-2xys}\right)}{\sqrt{x^2+y^2-2xys}}ds\right)\Phi_\kappa(t)dt \\ &= -2yb_\kappa\int_{0}^{1}\left(\int_{-t}^{t}\frac{f'\left(\sqrt{x^2+y^2-2xys}\right)}{\sqrt{x^2+y^2-2xys}}ds\right)t(1-t^2)^{\kappa-1}dt \\ &= -yb_\kappa\int_{-1}^{1}\frac{f'\left(\sqrt{x^2+y^2-2xys}\right)}{\sqrt{x^2+y^2-2xys}}\left(\int_{|s|}^{1}t(1-t^2)^{\kappa-1}dt\right)ds\end{aligned}$$

$$= -\frac{y}{\kappa}\int_{-1}^{1}\frac{f'\left(\sqrt{x^2+y^2-2xys}\right)}{\sqrt{x^2+y^2-2xys}}(1-s)\Phi_\kappa(s)ds.$$

Together with the formula for $\tau_y f'(x)$, this leads to

$$\mathscr{D}\tau_y f(x) = \int_{-1}^{1} f'\left(\sqrt{x^2+y^2-2xys}\right)\frac{x-y}{\sqrt{x^2+y^2-2xys}}\Phi_\kappa(s)ds.$$

Consequently, replacing f' by any odd function f_o, we conclude that

$$\tau_y f_o(x) = \int_{-1}^{1} f_o\left(\sqrt{x^2+y^2-2xys}\right)\frac{x-y}{\sqrt{x^2+y^2-2xys}}\Phi_\kappa(s)ds.$$

Any function f can be written as $f = f_e + f_o$, where $f_e(x) = (f(x)+f(-x))/2$ is the even part and $f_o(x) = (f(x)-f(-x))/2$ is the odd part, from which the stated formula follows. □

The explicit formula readily extends to the case of $G = \mathbb{Z}_2^d$ and the product weight function.

Theorem 6.3.8. *For $G = \mathbb{Z}_2^d$ and $h_\kappa(x) = \prod_{i=1}^d |x_i|^{\kappa_i}$ on $\mathbb{R}^d$,*

$$\tau_y f(x) = \tau_{y_1}\cdots\tau_{y_d} f(x), \qquad y = (y_1,\ldots,y_d)\in\mathbb{R}^d.$$

Proof. For $G = \mathbb{Z}_2^d$, the explicit formula of V_κ in (2.3.2) shows that

$$E(ix,y) = E(ix_1,y_1)\cdots E(ix_d,y_d)$$

for $x,y\in\mathbb{R}^d$, from which the theorem follows upon taking the Dunkl transform of τ_y. □

Note that the explicit formula also shows that $\tau_y f$ is not a positive operator. For example, we have

$$\tau_y(x_1-x_2)^2 = [(x_1-y_1)-(x_2-y_2)]^2 + \frac{4\kappa_1}{2\kappa_1}x_1y_1 + \frac{4\kappa_2}{2\kappa_2}x_2y_2.$$

Choosing $x_1 = -y_1 = 1$ and $x_2 = -y_2 = 1$ shows that $\tau_y(x_1-x_2)^2$ is not positive.

Using the formula for τ_y, we can establish the boundedness of the translation operator in the case of $\mathbb{Z}_2^d$.

Theorem 6.3.9. *For each $y\in\mathbb{R}^d$, the generalized translation operator τ_y is a bounded operator on $L^p(\mathbb{R}^d, h_\kappa^2)$. More precisely, $\|\tau_y f\|_{\kappa,p} \le 3\|f\|_{\kappa,p}$, $1\le p\le\infty$.*

Proof. The product nature of τ_y and h_κ means that we only have to consider the case $d = 1$. We have

$$\left|\int_{-1}^{1} f\left(\sqrt{x^2+y^2-2xyt}\right)\left(1+\frac{x-y}{\sqrt{x^2+y^2-2xyt}}\right)\Phi_\kappa(t)dt\right|$$
$$\le \|f\|_\infty + c_\kappa\int_{-1}^{1}\left|f\left(\sqrt{x^2+y^2-2xyt}\right)\right|\frac{|x-y|}{\sqrt{x^2+y^2-2xyt}}(1+t)(1-t^2)^{\kappa-1}dt.$$

Since $(x-y)(1+t) = (x-yt)-(y-xt)$, it follows that

$$\frac{|x-y|}{\sqrt{x^2+y^2-2xyt}}(1+t) \le 2.$$

Consequently, the above integral is bounded by $3\|f\|_\infty$. Hence, by the explicit formula of $\tau_y f$, we get $\|\tau_y f\|_\infty \le 3\|f\|_\infty$.

Next we consider the case $p=1$. For $f \in L^1(\mathbb{R}^d, h_\kappa^2)$ the mapping $g \mapsto Lg$ defined by $Lg = c_h \int_{\mathbb{R}^d} \tau_y f(x) g(x) h_\kappa^2(x) dx$ is a linear functional L on $C_0(\mathbb{R}^d)$. Using the property (1) of Proposition 6.3.1 we get

$$\begin{aligned}\left|\int_{\mathbb{R}^d} g(x)\tau_y f(x) h_\kappa^2(x)dx\right| &= \left|\int_{\mathbb{R}^d} \tau_{-y} g(x) f(x) h_\kappa^2(x)dx\right| \\ &\le \|\tau_y g\|_\infty \|f\|_{\kappa,1} \le 3\|g\|_\infty \|f\|_{\kappa,1},\end{aligned}$$

where we have used the fact that $\|\tau_y g\|_\infty \le 3\|g\|_\infty$. Hence, L is a bounded linear functional on $C_0(\mathbb{R}^d)$. By the Riesz representation theorem, $\tau_y f(x)dx$ is the unique regular measure providing the integral representation of L. Consequently,

$$\|\tau_y f\|_{\kappa,1} = \sup_{\|g\|_\infty = 1} \left|\int_{\mathbb{R}^d} g(x) f(x) h_\kappa^2(x)dx\right| \le 3\|f\|_{\kappa,1}.$$

Finally, interpolation shows that the same holds for $1 < p < \infty$. □

6.4 Generalized convolution and summability

The convolution $f *_\kappa g$ in Definition 6.2.8 is defined for $f, g \in L^2(\mathbb{R}^d; h_\kappa^2)$. By (6.2.4), it satisfies the relations

$$\mathscr{F}_\kappa(f *_k g) = \mathscr{F}_\kappa f \cdot \mathscr{F}_\kappa g \quad \text{and} \quad f *_\kappa g = g *_\kappa f. \tag{6.4.1}$$

6.4.1 Convolution with radial functions

If $g \in L^1(\mathbb{R}^d; h_\kappa^2)$, then $\mathscr{F}_\kappa g$ is bounded so that, by the Plancherel theorem,

$$\|f *_\kappa g\|_{\kappa,2} \le \|\mathscr{F}_\kappa g\|_\infty \|f\|_{k,2} \le \|g\|_{\kappa,1} \|f\|_{k,2}.$$

Since we do not know if the generalized translation operator is bounded in $L^p(\mathbb{R}^d; h_\kappa^2)$, the usual proof of Young's inequality does not apply. For convolution with radial functions we can state the following theorem.

Theorem 6.4.1. *Let g be a bounded radial function in $L^1(\mathbb{R}^d; h_\kappa^2)$. Then the map $f \mapsto f *_\kappa g$ extends to all $L^p(\mathbb{R}^d; h_\kappa^2)$, $1 \le p \le \infty$, as a bounded operator. In particular,*

$$\|f *_\kappa g\|_{\kappa,p} \le \|g\|_{\kappa,1} \|f\|_{\kappa,p}. \tag{6.4.2}$$

Proof. For $g \in L^1(\mathbb{R}^d; h_\kappa^2)$, bounded and radial, we have $|\tau_y g| \le \tau_y |g|$, which shows that

$$\int_{\mathbb{R}^d} |\tau_y g(x)| h_\kappa^2(x) dx \le \int_{\mathbb{R}^d} |g(x)| h_\kappa^2(x) dx.$$

Therefore,

$$\int_{\mathbb{R}^d} |f *_\kappa g(x)| h_\kappa^2(x) dx \le \|f\|_{\kappa,1} \|g\|_{\kappa,1}.$$

We also have $\|f *_\kappa g\|_\infty \le \|f\|_\infty \|g\|_{\kappa,1}$. By the Riesz–Thorin interpolation theorem, we obtain $\|f *_\kappa g\|_{\kappa,p} \le \|g\|_{\kappa,1} \|f\|_{\kappa,p}$. □

For $\phi \in L^1(\mathbb{R}^d; h_\kappa^2)$ and $\varepsilon > 0$, we define the dilation ϕ_ε by

$$\phi_\varepsilon(x) = \varepsilon^{-(2\gamma_\kappa + d)} \phi(x/\varepsilon) = \varepsilon^{-(2\lambda_\kappa + 2)} \phi(x/\varepsilon).$$

A change of variables shows that

$$\int_{\mathbb{R}^d} \phi_\varepsilon(x) h_\kappa^2(x) dx = \int_{\mathbb{R}^d} \phi(x) h_\kappa^2(x) dx, \qquad \text{for all } \varepsilon > 0.$$

Theorem 6.4.2. *Let $\phi \in L^1(\mathbb{R}^d; h_\kappa^2)$ be a bounded radial function and assume that it satisfies $c_h \int_{\mathbb{R}^d} \phi(x) h_\kappa^2(x) dx = 1$. Then for $f \in L^p(\mathbb{R}^d; h_\kappa^2)$, $1 \le p < \infty$, and $f \in C_0(\mathbb{R}^d)$, $p = \infty$,*

$$\lim_{\varepsilon \to 0} \|f *_\kappa \phi_\varepsilon - f\|_{\kappa,p} = 0.$$

Proof. For a given $\eta > 0$ we choose $g \in C_0^\infty$ such that $\|g - f\|_{\kappa,p} < \eta/3$. The triangle inequality and (6.4.2) lead to

$$\|f *_\kappa \phi_\varepsilon - f\|_{\kappa,p} \le \frac{2}{3}\eta + \|g *_\kappa \phi_\varepsilon - g\|_{\kappa,p}.$$

Since ϕ is radial, we can choose a radial function $\psi \in C_0^\infty$ such that

$$\|\phi - \psi\|_{\kappa,1} \le (12\|g\|_{\kappa,p})^{-1} \eta.$$

If we let $a = c_h \int_{\mathbb{R}^d} \psi(y) h_\kappa^2(y) dy$, then, by the triangle inequality and (6.4.2),

$$\begin{aligned} \|g *_\kappa \phi_\varepsilon - g\|_{\kappa,p} &\le \|g\|_{\kappa,p} \|\phi - \psi\|_{\kappa,1} + \|g *_\kappa \psi_\varepsilon - ag\|_{\kappa,p} + |a-1| \|g\|_{\kappa,p} \\ &\le \eta/6 + \|g *_\kappa \psi_\varepsilon - ag\|_{\kappa,p}, \end{aligned}$$

since $\|g\|_{\kappa,p} \|\phi - \psi\|_{\kappa,1} \le \frac{\eta}{12}$ and

$$|a-1| = \left| c_h \int_{\mathbb{R}^d} (\phi_\varepsilon(x) - \psi_\varepsilon(x)) h_\kappa^2(x) dx \right| \le (12\|g\|_{\kappa,p})^{-1} \eta.$$

Thus,

$$\|f *_\kappa \phi_\varepsilon - f\|_{\kappa,p} \le \frac{5}{6}\eta + \|g *_\kappa \psi_\varepsilon - ag\|_{\kappa,p}.$$

Hence it suffices to show that $\|g *_\kappa \psi_\varepsilon - ag\|_{\kappa,p} \le \eta/6$. Now $g \in A_\kappa(\mathbb{R}^d)$, hence

$$g *_\kappa \phi_\varepsilon(x) = \int_{\mathbb{R}^d} g(y)\tau_x\widetilde{\phi_\varepsilon}(y)h_\kappa^2(y)dy = \int_{\mathbb{R}^d} \tau_{-x}g(y)\phi_\varepsilon(-y)h_\kappa^2(y)dy.$$

We also know that $\tau_{-x}g(y) = \tau_{-y}g(x)$, as $g \in C_0^\infty$. Therefore,

$$g *_\kappa \phi_\varepsilon(x) = \int_{\mathbb{R}^d} \tau_y g(x)\phi_\varepsilon(y)h_\kappa^2(y)dy.$$

In view of this

$$g *_\kappa \psi_\varepsilon(x) - ag(x) = \int_{\mathbb{R}^d} (\tau_y g(x) - g(x))\,\psi_\varepsilon(y)h_\kappa^2(y)dy,$$

which gives, by Minkowski's integral inequality,

$$\|g *_\kappa \psi_\varepsilon - ag\|_{\kappa,p} \le \int_{\mathbb{R}^d} \|\tau_y g - g\|_{\kappa,p}\psi_\varepsilon(y)|h_\kappa^2(y)dy.$$

Now, using Theorem 6.2.7, the rest of the proof follows as in the proof of Lemma 6.2.9. □

6.4.2 Summability of the inverse Dunkl transform

With Theorem 6.4.2 established, we can now extend our proof of the inversion formula in Theorem 6.2.10 to a more general summability method of the inverse Dunkl transform.

Let $\Phi \in L^1(\mathbb{R}^d; h_\kappa^2)$ be continuous at 0 and assume $\Phi(0) = 1$. For $f \in \mathscr{S}$ and $\varepsilon > 0$ define

$$T_\varepsilon f(x) = c_h \int_{\mathbb{R}^d} \mathscr{F}_\kappa f(y)E(ix,y)\Phi(-\varepsilon y)h_\kappa^2(y)dy.$$

It is clear that T_ε extends to the whole of L^2 as a bounded operator: this follows from Plancherel's theorem. Let us study the convergence of $T_\varepsilon f$ as $\varepsilon \to 0$. Note that $T_0 f = f$ by the inversion formula for the Dunkl transform.

If $T_\varepsilon f$ can be extended to all $f \in L^p(\mathbb{R}^d; h_\kappa^2)$ and if $T_\varepsilon f \to f$ in $L^p(\mathbb{R}^d; h_\kappa^2)$, we say that the inverse Dunkl transform is Φ-summable.

Proposition 6.4.3. *Suppose both Φ and $\phi = \mathscr{F}_\kappa \Phi$ belong to $L^1(\mathbb{R}^d; h_\kappa^2)$. If Φ is radial, then*

$$T_\varepsilon f(x) = (f *_\kappa \phi_\varepsilon)(x)$$

for all $f \in L^2(\mathbb{R}^d; h_\kappa^2)$ and $\varepsilon > 0$.

Proof. Under the hypothesis on Φ, both T_ε and the operator taking f into $(f *_\kappa \phi_\varepsilon)$ extend to $L^2(\mathbb{R}^d; h_\kappa^2)$ as bounded operators. So it is enough to verify $T_\varepsilon f(x) = (f *_\kappa \phi_\varepsilon)(x)$ for all f in the Schwartz class. By the definition of the Dunkl transform,

$$
\begin{aligned}
T_\varepsilon f(x) &= c_h \int_{\mathbb{R}^d} \mathscr{F}_\kappa \tau_{-x} f(y) \Phi(-\varepsilon y) h_\kappa^2(y) dy \\
&= c_h \int_{\mathbb{R}^d} \tau_{-x} f(\xi) c_h \int_{\mathbb{R}^d} \Phi(-\varepsilon y) E(y, -i\xi) h_\kappa^2(y) dy h_\kappa^2(\xi) d\xi \\
&= c_h \varepsilon^{-(d+2\gamma_\kappa)} \int_{\mathbb{R}^d} \tau_{-x} f(\xi) \mathscr{F}_\kappa \Phi(-\varepsilon^{-1}\xi) h_\kappa^2(\xi) d\xi \\
&= (f *_\kappa \phi_\varepsilon)(x),
\end{aligned}
$$

where we have changed variable $\xi \mapsto -\xi$ and used the fact that $\tau_{-x} f(-\xi) = \tau_\xi f(x)$. $\square$

Theorem 6.4.4. *Let $\Phi(x) \in L^1(\mathbb{R}^d; h_\kappa^2)$ be radial and assume that $\mathscr{F}_\kappa \Phi \in L^1(\mathbb{R}^d; h_\kappa^2)$ is bounded and $\Phi(0) = 1$. For $f \in L^p(\mathbb{R}^d; h_\kappa^2)$, $T_\varepsilon f$ converges to f in $L^p(\mathbb{R}^d; h_\kappa^2)$ as $\varepsilon \to 0$, for $1 \le p < \infty$.*

Proof. Since $f *_\kappa \phi_\varepsilon$ agrees with $T_\varepsilon f$ on $L^2(\mathbb{R}^d; h_\kappa^2)$ by the previous theorem, and $f *_\kappa \phi_\varepsilon$ is bounded in $L^p(\mathbb{R}^d; h_\kappa^2)$, $T_\varepsilon f$ can be extended to $L^p(\mathbb{R}^d; h_\kappa^2)$. The convergence of $T_\varepsilon f$ to f now follows from Theorem 6.4.2. $\square$

By choosing specific radial functions Φ, we consider several examples of summability methods.

Heat kernel transform. We consider $f *_\kappa q_t^\kappa$, where q_t^κ is the heat kernel defined in (6.1.3). Recall the Dunkl Laplacian Δ_h defined in (2.2.3).

Theorem 6.4.5. *Suppose $f \in L^p(\mathbb{R}^d; h_\kappa^2)$, $1 \le p < \infty$ or $f \in C_0(\mathbb{R}^d)$, $p = \infty$.*

1. *The heat transform*

$$
H_t f(x) := (f *_\kappa q_t^\kappa)(x) = c_h \int_{\mathbb{R}^d} f(y) \tau_y q_t^\kappa(x) h_\kappa^2(y) dy, \qquad t > 0,
$$

converges to f in $L^p(\mathbb{R}^d; h_\kappa^2)$ as $t \to 0$.

2. *Define $H_0 f(x) = f(x)$. Then the function $H_t f(x)$ solves the initial value problem*

$$
\Delta_h u(x,t) = \partial_t u(x,t), \qquad u(x,0) = f(x), \qquad (x,t) \in \mathbb{R}^d \times [0,\infty).
$$

Proof. By (6.1.4), $\mathscr{F}_\kappa q_t^\kappa = e^{-t\|x\|^2}$, so that (1) follows by setting $\Phi(x) = e^{-\|x\|^2}$ and $\varepsilon = \sqrt{t}$. Using the spherical-polar form of Δ_h in (3.1.5), it is easy to see that q_t^κ satisfies the heat equation $\Delta_h u(x,t) = \partial_t u(x,t)$, so that (2) holds. $\square$

Poisson integral. The Poisson kernel is defined by

$$
P(x,\varepsilon) := c_{d,\kappa} \frac{\varepsilon}{(\varepsilon^2 + \|x\|^2)^{\lambda_\kappa + \frac{3}{2}}}, \qquad c_{d,\kappa} = 2^{\lambda_\kappa + 1} \frac{\Gamma(\lambda_\kappa + \frac{3}{2})}{\sqrt{\pi}}. \tag{6.4.3}
$$

Theorem 6.4.6. *Suppose $f \in L^p(\mathbb{R}^d;h_\kappa^2)$, $1 \le p < \infty$, or $f \in C_0(\mathbb{R}^d)$, $p = \infty$. Then the Poisson integral $f *_\kappa P_\varepsilon$ converges to f in $L^p(\mathbb{R}^d;h_\kappa^2)$.*

Proof. Set $\Phi(x) = e^{-\|x\|}$. Then $P(x,1) = \mathscr{F}_\kappa\Phi(x)$ and $P(x,\varepsilon) = \varepsilon^{-\lambda_\kappa-2}P(x/\varepsilon,1)$. Indeed, this can be proved as for the ordinary Fourier transform using the integral relation between e^{-t} and e^{-t^2} in (3.4.8) and the fact that the Dunkl transform of $e^{-\|x\|^2/4}$ is $e^{-\|x\|^2}$, as shown by (6.1.4). Since $\Phi(0) = 1$, it readily follows that $c_h \int_{\mathbb{R}^d} P(x,\varepsilon)h_\kappa^2(x)dx = 1$. Thus, the convergence is established by Theorem 6.4.4. □

Bochner–Riesz means. Here we consider

$$\Phi(x) = \begin{cases} (1-\|x\|^2)^\delta, & \|x\| \le 1, \\ 0, & \text{otherwise}, \end{cases}$$

where $\delta > 0$. As in the case of the ordinary Fourier transform, we take $\varepsilon = 1/R$ where $R > 0$. Then the Bochner–Riesz means of order δ are defined by

$$S_R^\delta f(x) := c_h \int_{\|y\| \le R} \left(1 - \frac{\|y\|}{R}\right)^\delta \mathscr{F}_\kappa f(y)E(ix,y)h_\kappa^2(y)dy. \tag{6.4.4}$$

Theorem 6.4.7. *If $f \in L^p(\mathbb{R}^d;h_\kappa^2)$, $1 \le p \le \infty$ and $\delta > \frac{d-1}{2} + \gamma_\kappa$, then*

$$\|S_R^\delta f - f\|_{\kappa,p} \to 0, \qquad \text{as } R \to \infty.$$

Proof. The proof follows as in the case of ordinary Fourier transform [52, p. 171]. From Theorem 6.1.8 and the properties of the Bessel function, we have

$$\mathscr{F}_\kappa\Phi(x) = 2^{\lambda_\kappa}\|x\|^{-\lambda_\kappa-\delta-1}J_{\lambda_\kappa+\delta+1}(\|x\|).$$

It is known that $J_\alpha(r) = O(r^{-1/2})$, so that $\mathscr{F}_\kappa\Phi \in L^1(\mathbb{R}^d,h_\kappa^2)$ under the condition $\delta > \lambda_\kappa + 1/2$. □

We note that $\lambda_\kappa = (d-2)/2 + \gamma_\kappa$, where γ_κ is the sum of all (nonnegative) parameters in the weight function. If all parameters are zero, then $h_\kappa(x) \equiv 1$ and we are back to the classical Fourier transform, for which the index $(d-1)/2$ is the critical index for the Bochner–Riesz means.

6.4.3 Convolution operator for $\mathbb{Z}_2^d$

In the case of $h_\kappa(x)$ associated with the group $\mathbb{Z}_2^d$, the translation operator τ_y is bounded on $L^p(\mathbb{R}^d,h_\kappa^2)$. Hence, the standard proof can be used to establish the following inequality:

Theorem 6.4.8. *Let $G = \mathbb{Z}_2^d$. Let $p,q,r \ge 1$ and $p^{-1} = q^{-1} + r^{-1} - 1$. Let $f \in L^q(\mathbb{R}^d,h_\kappa^2)$ and $g \in L^r(\mathbb{R}^d,h_\kappa^2)$. Then*

$$\|f *_\kappa g\|_{\kappa,p} \le c\|f\|_{\kappa,q}\|g\|_{\kappa,r}.$$

The boundedness of τ_y allows us to remove the assumption that ϕ is radial in Theorem 6.4.2 when $G = \mathbb{Z}_2^d$.

Theorem 6.4.9. *Let $\phi \in L^1(\mathbb{R}^d, h_\kappa^2)$ with $\int_{\mathbb{R}^d} \phi(x) h_\kappa^2(x) dx = 1$. Then for $f \in L^p(\mathbb{R}^d, h_\kappa^2)$ if $1 \le p < \infty$, or $f \in C_0(\mathbb{R}^d)$ if $p = \infty$,*

$$\lim_{\varepsilon \to 0} \|f *_\kappa \phi_\varepsilon - f\|_{\kappa,p} = 0, \qquad 1 \le p \le \infty.$$

Proof. For $f \in L^p(\mathbb{R}^d, h_\kappa^2)$ we write $f = f_1 + f_2$, where f_1 is in C_0^∞ with compact support, and $\|f_2\|_{\kappa,p} \le \delta$. Then the second term of the inequality

$$\|\tau_y f(x) - f(x)\|_{\kappa,p} \le \|\tau_y f_1(x) - f_1(x)\|_{\kappa,p} + \|\tau_y f_2(x) - f_2(x)\|_{\kappa,p}$$

is bounded by $(1+c)\delta$, as τ_y is a bounded operator, and the first term goes to zero as $\varepsilon \to 0$ by Theorem 6.2.7. This proves that $\|\tau_y f(x) - f(x)\|_{\kappa,p} \to 0$ as $y \to 0$. We have then

$$\begin{aligned}
& c_h \int_{\mathbb{R}^d} |f *_\kappa g_\varepsilon(x) - f(x)|^p h_\kappa^2(x) dx \\
&= c_h \int_{\mathbb{R}^d} \left| c_h \int_{\mathbb{R}^d} (\tau_y f(x) - f(x)) g_\varepsilon(y) h_\kappa^2(y) dy \right|^p h_\kappa^2(x) dx \\
&\le c_h \int_{\mathbb{R}^d} \|\tau_y f - f\|_{\kappa,p}^p |g_\varepsilon(x)| h_\kappa^2(x) dx \\
&= c_h \int_{\mathbb{R}^d} \|\tau_{\varepsilon y} f - f\|_{\kappa,p}^p |g(x)| h_\kappa^2(x) dx,
\end{aligned}$$

which goes to zero as $\varepsilon \to 0$. □

All results on the summability for the inverse Dunkl transform in the previous section are proved for a generic reflection group, hence they all hold for the case of $\mathbb{Z}_2^d$.

6.5 Maximal function

We study the Hardy–Littlewood maximal function in weighted spaces.

6.5.1 Boundedness of maximal function

Let $B_r = B(0,r)$ denote the ball of radius r centered at the origin, and let χ_{B_r} denote its characteristic function.

Definition 6.5.1. For $f \in L^1(\mathbb{R}^d; h_\kappa^2)$, we define the maximal function $M_\kappa f$ by

$$M_\kappa f(x) := \sup_{r>0} \frac{\left|\int_{\mathbb{R}^d} f(y) \tau_x \chi_{B_r}(y) h_\kappa^2(y) dy\right|}{\int_{B_r} h_\kappa^2(y) dy}.$$

The function χ_{B_r} is radial. If $\varphi \in C_0^\infty(\mathbb{R}^d)$ is a radial function such that $\chi_{B_r}(x) \le \varphi(x)$, then, by Theorem 6.3.3, $\tau_y \chi_{B_r}(x) \le \tau_y \varphi(x)$. As $\tau_y \varphi$ is bounded, $\tau_y \chi_{B_r}$ is bounded and compactly supported so that it belongs to $L^1(\mathbb{R}^d; h_\kappa^2)$. Hence, $M_\kappa f$ is well defined for $f \in L^1(\mathbb{R}^d; h_\kappa^2)$. Using spherical polar coordinates, we also have

$$\int_{B_r} h_\kappa^2(y)dy = \int_0^r s^{d-1+2\gamma_\kappa} ds \int_{\mathbb{S}^{d-1}} h_\kappa^2(\xi) d\sigma(\xi) = \frac{\omega_d^\kappa}{2\lambda_\kappa + 2} r^{2\lambda_\kappa+2}.$$

By definition, we can also write $M_\kappa f$ as

$$M_\kappa f(x) = \sup_{r>0} \frac{1}{d_\kappa r^{2\lambda_\kappa+2}} |f *_\kappa \chi_{B_r}(x)|, \quad d_\kappa := \frac{\omega_d^\kappa}{2\lambda_\kappa + 2}.$$

Since $\tau_y \chi_{B_r} \ge 0$, we have $M_\kappa f(x) \le M_\kappa |f|(x)$.

Theorem 6.5.2. *The maximal function is bounded on $L^p(\mathbb{R}^d; h_\kappa^2)$ for $1 < p \le \infty$; moreover it is of weak type $(1,1)$, that is, for $f \in L^1(\mathbb{R}^d; h_\kappa^2)$ and $\alpha > 0$,*

$$\int_{E(a)} h_\kappa^2(x)dx \le \frac{c}{\alpha} \|f\|_{\kappa,1},$$

where $E(\alpha) = \{x : M_\kappa f(x) > \alpha\}$ and c is a constant independent of α and f.

Proof. Without loss of generality we can assume that $f \ge 0$. Let $P_\varepsilon(x) := P(x, \varepsilon)$ be the Poisson kernel defined in (6.4.3), and let $\sigma := 2\lambda_\kappa + 3$. For $j \ge 0$, define $B_{r,j} := \{x : 2^{-j-1} r \le \|x\| \le 2^{-j} r\}$. Then

$$\begin{aligned}\chi_{B_{r,j}}(y) &= (2^{-j}r)^\sigma (2^{-j}r)^{-\sigma} \chi_{B_{r,j}}(y)\\ &\le c(2^{-j}r)^{\sigma-1} \frac{2^{-j}r}{((2^{-j}r)^2 + \|y\|^2)^{\sigma/2}} \chi_{B_{r,j}}(y)\\ &\le c(2^{-j}r)^{\sigma-1} P_{2^{-j}r}(y),\end{aligned}$$

where c is a constant independent of r and j. Since both χ_{B_r} and P_ε are bounded, integrable radial functions, it follows from Theorem 6.3.3 that

$$\tau_x \chi_{B_{r,j}}(y) \le c(2^{-j}r)^{\sigma-1} \tau_x P_{2^{-j}r}(y).$$

This shows that for any positive integer m

$$\begin{aligned}\int_{\mathbb{R}^d} f(y) \sum_{j=0}^m \tau_x \chi_{B_{r,j}}(y) h_\kappa^2(y)dy &\le c \sum_{j=0}^\infty (2^{-j}r)^{\sigma-1} \int_{\mathbb{R}^d} f(y) \tau_x P_{2^{-j}r}(y) h_\kappa^2(y)dy\\ &\le c\, r^{d+2\gamma_\kappa} \sup_{t>0} f *_\kappa P_t(x).\end{aligned}$$

As $\sum_{j=0}^m \chi_{B_{r,j}}(y)$ converges to $\chi_{B_r}(y)$ in $L^1(\mathbb{R}^d; h_\kappa^2)$, the boundedness of τ_x on $L^1_{\mathrm{rad}}(\mathbb{R}^d; h_\kappa^2)$ shows that $\sum_{j=0}^m \tau_x \chi_{B_{r,j}}(y)$ converges to $\tau_x \chi_{B_r}(y)$ in $L^1(\mathbb{R}^d; h_\kappa^2)$. By passing to a subsequence if necessary, we can assume that $\sum_{j=0}^m \tau_x \chi_{B_{r,j}}(y)$ converges to $\tau_x \chi_{B_r}(y)$ for almost every y. Thus all the functions involved are uniformly bounded by $\tau_x\chi_{B_r}(y)$. This shows that $\sum_{j=0}^m \tau_x\chi_{B_{r,j}}(y)$ converges to $\tau_x\chi_{B_r}(y)$ in $L^{p'}(\mathbb{R}^d;h_\kappa^2)$ and hence

$$\lim_{m\to\infty}\int_{\mathbb{R}^d} f(y)\sum_{j=0}^m \tau_x\chi_{B_{r,j}}(y)h_\kappa^2(y)dy = \int_{\mathbb{R}^d} f(y)\tau_x\chi_{B_r}(y)h_\kappa^2(y)dy.$$

Thus we have proved

$$f *_\kappa \chi_{B_r}(x) \le cr^{d+2\gamma_\kappa}\sup_{t>0} f *_\kappa P_t(x),$$

which gives the inequality $M_\kappa f(x) \le cP^*f(x)$, where $P^*f(x) = \sup_{t>0} f *_\kappa P_t(x)$ is the maximal function associated to the Poisson semi-group.

Therefore, it is enough to prove the boundedness of P^*f. By looking at the Dunkl transforms of the Poisson kernel and the heat kernel we conclude, as in the proof of Lemma 3.4.7, that

$$f *_\kappa P_t(x) = \frac{t}{\sqrt{2\pi}}\int_0^\infty (f *_\kappa q_s)(x)e^{-t^2/2s}s^{-3/2}ds,$$

which allows us to conclude that

$$P^*f(x) \le c\sup_{t>0}\frac{1}{t}\int_0^t Q_s(|f|)(x)ds,$$

where $Q_sf(x) = f *_\kappa q_s(x)$ is the heat semi-group. Hence, using the Hopf–Dunford–Schwartz ergodic theorem (Theorem 3.4.3), we conclude the boundedness of P^*f on $L^p(\mathbb{R}^d;h_\kappa^2)$ for $1 < p \le \infty$, and the weak type $(1,1)$. □

The maximal function can be used to study almost everywhere convergence of the convolution $f *_\kappa \varphi_\varepsilon$ when ϕ satisfies moderate conditions.

Theorem 6.5.3. *Let $\phi \in A_\kappa(\mathbb{R}^d)$ be a real valued radial function which satisfies $|\phi(x)| \le c(1+\|x\|)^{-2\lambda_\kappa-3}$. Then*

$$\sup_{\varepsilon>0}|f *_\kappa \phi_\varepsilon(x)| \le cM_\kappa f(x).$$

Consequently, $f *_\kappa \phi_\varepsilon(x) \to f(x)$ for almost every x as ε goes to 0, for all f in $L^p(\mathbb{R}^d;h_\kappa^2)$, $1 \le p < \infty$.

Proof. We can assume that both f and ϕ are nonnegative. Writing

$$\phi_\varepsilon(y) = \sum_{j=-\infty}^{\infty} \phi_\varepsilon(y)\chi_{\varepsilon 2^j \le \|y\| \le \varepsilon 2^{j+1}}(y),$$

we have

$$\left|\tau_x\left[\sum_{j=-m}^m \phi_\varepsilon\chi_{\varepsilon 2^j \le \|y\| \le \varepsilon 2^{j+1}}\right](y)\right| \le c\sum_{j=-m}^m (1+2^j)^{-2\lambda_\kappa-3}\varepsilon^{-2\lambda_\kappa-2}.$$

This shows that

$$\int_{\mathbb{R}^d} f(y)\tau_x\left[\phi_\varepsilon(y)\sum_{j=-m}^{m}\chi_{\varepsilon 2^j\le\|y\|\le\varepsilon 2^{j+1}}\right](y)h_\kappa^2(y)dy$$
$$\le c\varepsilon^{-2\lambda_\kappa-2}\sum_{j=-m}^{m}(1+2^j)^{-2\lambda_\kappa-3}(\varepsilon 2^j)^{2\lambda_\kappa+2}M_\kappa f(x)\le cM_\kappa f(x).$$

Since $|\phi(y)|\le c(1+\|y\|)^{-2\lambda_\kappa-3}\le cP_1(y)$ it follows that $|\tau_x\phi(y)|\le c\tau_x P_1(y)$ is bounded. Arguing as in the previous theorem, we can show that the left-hand side of the above inequality converges to $f*_\kappa\phi_\varepsilon(x)$. Thus we obtain

$$\sup_{\varepsilon>0}|f*_\kappa\phi_\varepsilon(x)|\le cM_\kappa f(x),$$

from which the proof of the almost everywhere convergence follows via the standard argument. □

6.5.2 Convolution versus maximal function for $\mathbb{Z}_2^d$

In the case of $\mathbb{Z}_2^d$, the conditions of the last theorem can be relaxed. For this, we need the spherical mean operator defined on $A_\kappa(\mathbb{R}^d)$ by

$$S_r f(x) := \frac{1}{\omega_d^\kappa}\int_{S^{d-1}}\tau_{ry}f(x)h_\kappa^2(y)d\sigma(y).$$

If $f\in A_\kappa(\mathbb{R}^d)$ and $g(x)=g_0(\|x\|)$ is an integrable radial function, then, using spherical-polar coordinates, the generalized convolution $f*_\kappa g$ can be expressed in terms of the spherical mean:

$$\begin{aligned}(f*_\kappa g)(x) &= c_h\int_{\mathbb{R}^d}\tau_y f(x)g(y)h_\kappa^2(y)dy\\ &= c_h\int_0^\infty r^{2\lambda_\kappa+1}g_0(r)\left(\int_{S^{d-1}}\tau_{ry'}f(x)h_\kappa^2(y')dy'\right)dr\\ &= c_h\omega_d^\kappa\int_0^\infty S_r f(x)g_0(r)r^{2\lambda_\kappa+1}dr.\end{aligned}$$

The spherical mean operator is bounded.

Theorem 6.5.4. *Let $G=\mathbb{Z}_2^d$. For $f\in L^p(\mathbb{R}^d,h_\kappa^2)$,*

$$\|S_r f\|_{\kappa,p}\le c\|f\|_{\kappa,p},\qquad 1\le p\le\infty.$$

Furthermore, $\|S_r f-f\|_{\kappa,p}\to 0$ as $r\to 0^+$.

Proof. Using Hölder's inequality,

$$|S_r f(x)|^p\le\frac{1}{\omega_d^\kappa}\int_{S^{d-1}}|\tau_{ry}f(x)|^p h_\kappa^2(y)d\sigma(y).$$

Hence, a simple computation shows that

$$\begin{aligned} c_h \int_{\mathbb{R}^d} |S_r f(x)|^p h_\kappa^2(x)dx &\le c_h \int_{\mathbb{R}^d} \frac{1}{\omega_d^\kappa}\Big(\int_{S^{d-1}} |\tau_{ry} f(x)|^p h_\kappa^2(y)d\omega(y)\Big) h_\kappa^2(x)dx \\ &= \frac{1}{\omega_d^\kappa}\int_{S^{d-1}} \|\tau_{ry} f\|_{\kappa,p}^p h_\kappa^2(y)d\sigma(y) \\ &\le c\|f\|_{\kappa,p}. \end{aligned}$$

Furthermore,

$$\|S_r f - f\|_{\kappa,p}^p \le \frac{1}{\omega_d^\kappa}\int_{S^{d-1}} \|\tau_{ry} f - f\|_{\kappa,p}^p h_\kappa^2(y)d\sigma(y),$$

which goes to zero as $r \to 0$, since $\|\tau_{ry} f - f\|_{\kappa,p} \to 0$. □

Theorem 6.5.5. *Set $G = \mathbb{Z}_2^d$. Let $\phi(x) = \phi_0(\|x\|) \in L^1(\mathbb{R}^d; h_\kappa^2)$ be a radial function. Assume that ϕ_0 is differentiable, $\lim_{r\to\infty}\phi_0(r) = 0$, and $\int_0^\infty r^{2\lambda_\kappa+2}|\phi_0(r)|dr < \infty$. Then*

$$|(f *_\kappa \phi)(x)| \le cM_\kappa f(x).$$

In particular, if $\phi \in L^1(\mathbb{R}^d; h_\kappa^2)$ and $c_h \int_{\mathbb{R}^d}\phi(x)h_\kappa^2(x)dx = 1$, then

1. *for $1 \le p \le \infty$, $f *_\kappa \phi_\varepsilon$ converges to f as $\varepsilon \to 0$ in $L^p(\mathbb{R}^d; h_\kappa^2)$;*
2. *for $f \in L^1(\mathbb{R}^d, h_\kappa^2)$, $(f *_\kappa \phi_\varepsilon)(x)$ converges to $f(x)$ as $\varepsilon \to 0$ for almost all $x \in \mathbb{R}^d$.*

Proof. By definition of the spherical mean $S_t f$, we can also write

$$M_\kappa f(x) = \sup_{r>0} \frac{\left|\int_0^r t^{2\lambda_\kappa+1} S_t f(x)dt\right|}{\int_0^r t^{2\lambda_\kappa+1}dt}.$$

Since $|M_\kappa f(x)| \le cM_\kappa|f|(x)$, we can assume $f(x) \ge 0$. The assumption on ϕ_0 shows that

$$\begin{aligned} \lim_{r\to\infty} \phi_0(r)\int_0^r S_t f(x) t^{2\lambda_\kappa+1}dt &= \lim_{r\to\infty}\phi_0(r)\int_{\mathbb{R}^d} \tau_y f(x) h_\kappa^2(y)dy \\ &= \lim_{r\to\infty}\phi_0(r)\int_{\mathbb{R}^d} f(y)h_\kappa^2(y)dy = 0. \end{aligned}$$

Hence, using the spherical-polar coordinates and integrating by parts, we get

$$\begin{aligned} (f *_\kappa \phi)(x) &= \int_0^\infty \phi_0(r) r^{2\lambda_\kappa+1} S_r f(x)dr \\ &= -\int_0^\infty \Big(\int_0^r S_t f(x) t^{2\lambda_\kappa+1}dt\Big)\phi'(r)dr, \end{aligned}$$

which implies that

$$|(f *_\kappa \phi)(x)| \le cM_\kappa f(x)\int_0^\infty r^{2\lambda_\kappa+2}|\phi_0'(r)|dr.$$

Boundedness of the last integral proves the maximal inequality. □

We can further enhance Theorem 6.5.5 by removing the assumption that ϕ is radial. For this purpose, we make the following simple observation about the maximal function. If f is nonnegative, then we can drop the absolute value sign in the definition of the maximal function, even though $\tau_y f$ may not be nonnegative.

Lemma 6.5.6. *If $f \in L^1(\mathbb{R}^d, h_\kappa^2)$ is a nonnegative function, then*

$$M_\kappa f(x) = \sup_{r>0} \frac{\int_{B_r} \tau_y f(x) h_\kappa^2(y) dy}{\int_{B_r} h_\kappa^2(y) dy}.$$

In particular, if f and g are two nonnegative functions, then

$$M_\kappa f + M_\kappa g = M_\kappa(f+g).$$

Proof. Since $\tau_y \chi_{B_r}(x)$ is nonnegative, we have that

$$(f *_\kappa \chi_{B_r})(x) = \int_{\mathbb{R}^d} f(y) \tau_y \chi_{B_r}(x) h_\kappa^2(y) dy$$

is nonnegative if f is nonnegative. Hence, we can drop the absolute value symbol in the definition of $M_\kappa f$. □

Theorem 6.5.7. *Set $G = \mathbb{Z}_2^d$. Let $\phi \in L^1(\mathbb{R}^d, h_\kappa^2)$ and let $\psi(x) = \psi_0(\|x\|) \in L^1(\mathbb{R}^d, h_\kappa^2)$ be a nonnegative radial function such that $|\phi(x)| \le \psi(x)$. Assume that ψ_0 is differentiable, $\lim_{r\to\infty} \psi_0(r) = 0$, and $\int_0^\infty r^{2\lambda_\kappa+2} |\psi_0(r)| dr < \infty$. Then $\sup_{\varepsilon>0} |f *_\kappa \phi_\varepsilon(x)|$ is of weak type $(1,1)$. In particular, if $\phi \in L^1(\mathbb{R}^d, h_\kappa^2)$ and $c_h \int_{\mathbb{R}^d} \phi(x) h_\kappa^2(x) dx = 1$, then, for $f \in L^1(\mathbb{R}^d, h_\kappa^2)$, $(f *_\kappa \phi_\varepsilon)(x)$ converges to $f(x)$ as $\varepsilon \to 0$ for almost all $x \in \mathbb{R}^d$.*

Proof. Since $M_\kappa f(x) \le M_\kappa |f|(x)$, we can assume that $f(x) \ge 0$. The proof uses the explicit formula for $\tau_y f$. Let us first consider the case $d = 1$. Since ψ is an even function, $\tau_y \psi$ is given by

$$\tau_y f(x) = \int_{-1}^1 f\left(\sqrt{x^2+y^2-2xyt}\right) \Phi_\kappa(t) dt,$$

according to (6.3.2). Since $(x-y)(1+t) = (x-yt)-(y-xt)$, we have

$$\frac{|x-y|}{\sqrt{x^2+y^2-2xyt}}(1+t) \le 2.$$

Consequently, by the explicit formula for $\tau_y f$, see (6.3.2), the inequality $|\phi(x)| \le \psi(x)$ implies

$$|\tau_y \phi(x)| \le \tau_y \psi(x) + 2\tau_{y,1} \psi(x),$$

where $\tau_{y,1}\psi$ is defined by

$$\tau_{y,1}\psi(x) = b_\kappa \int_{-1}^1 f\left(\sqrt{x^2+y^2-2xyt}\right) (1-t^2)^{\kappa-1} dt.$$

Note that $\tau_{y,1}\psi$ differs from $\tau_y\psi$ by the factor $1+t$ in the weight function. The change of variables $t \mapsto -t$ and $y \mapsto -y$ in the integrals shows that

$$\int_{\mathbb{R}} f(y)\tau_{y,1}\psi(x)h_\kappa^2(y)dy = \int_{\mathbb{R}} F(y)\tau_y\psi(x)h_\kappa^2(y)dy,$$

where $F(y) = (f(y)+f(-y))/2$. It follows that

$$|(f *_\kappa \phi)(x)| = \left| c_h \int_{\mathbb{R}} f(y)\tau_y\phi(x)h_\kappa^2(y)dy \right| \le (f *_\kappa \psi)(x) + 2(F *_\kappa \psi)(x).$$

The same consideration can be extended to the case of $\mathbb{Z}_2^d$ for $d>1$. Let $\{e_1,\dots,e_d\}$ be the standard Euclidean basis. For $\delta_j = \pm 1$ define $x\delta_j = x-(1+\delta_j)x_je_j$ (that is, multiplying the j-th component of x by δ_j gives $x\delta_j$). For $1 \le j \le d$ we define

$$F_{j_1,\dots,j_k} = 2^{-k} \sum_{(\delta_{j_1},\dots,\delta_{j_k})\in\mathbb{Z}_2^k} f(x\delta_{j_1}\cdots\delta_{j_k}).$$

In particular,

$$F_j(x) = (F(x)+F(x\delta_j))/2, \quad F_{j_1,j_2}(x) = (F(x)+F(x\delta_{j_1})+F(x\delta_{j_2})+F(x\delta_{j_1}\delta_{j_2}))/4,$$

and the last sum is over $\mathbb{Z}_2^d$, $F_{1,\dots,d}(x) = 2^{-d}\sum_{\sigma\in\mathbb{Z}_2^d} f(x\sigma)$. Following the proof in the case $d=1$ it is not hard to see that

$$\begin{aligned}|(f *_\kappa \phi)(x)| \le (f *_\kappa \psi)(x) &+ 2\sum_{j=1}^d (F_j *_\kappa \psi)(x) + 4\sum_{j_1\ne j_2}(F_{j_1,j_2} *_\kappa \psi)(x)\\ &+\cdots+2^d(F_{1,\dots,d} *_\kappa \psi)(x).\end{aligned}$$

For $G=\mathbb{Z}_2^d$, the explicit formula of τ_y shows that $M_\kappa f(x)$ is even in each of its variables. Hence, applying the result of the previous theorem to each of the above terms, we get

$$\begin{aligned}|(f *_\kappa \phi)(x)| \le M_\kappa f(x) &+ 2\sum_{j=1}^d M_\kappa F_j(x) + 4\sum_{j_1\ne j_2} M_\kappa F_{j_1,j_2}(x)\\ &+\cdots+2^d M_\kappa F_{1,\dots,d}(x).\end{aligned}$$

Since all F_j are clearly nonnegative, by Lemma 6.5.6, the last expression can be written as $M_\kappa H$, where H is the sum of all functions involved. Consequently, since $\|F_{j_1,\dots,j_d}\|_{\kappa,1} \le \|f\|_{\kappa,1}$,

$$\int_{\{x:(f*_\kappa\phi)(x)\ge a\}} h_\kappa^2(y)dy \le c\frac{\|H\|_{\kappa,1}}{a} \le c_d\frac{\|f\|_{\kappa,1}}{a}.$$

Hence, $f *_\kappa \phi$ is of weak type $(1,1)$, from which the almost everywhere convergence follows as usual. □

6.6 Notes and further results

The Dunkl transform was introduced in [27], where the L^2 isometry was established as we have seen in Section 6.1. The Dunkl transform was studied in [34], where the bound for $E(x,y)$ was established for general parameters κ with $\mathrm{Re}\,\kappa \ge 0$ and without the positivity of V_κ when κ is real and nonnegative, and where the main results of the L^1 theory (Theorem 6.2.10) were also established. Our development in Section 6.2, based on the convolution operator, follows closely the approach of the classical harmonic analysis (see, for example, [52]). The generalized translation operator was studied in [39, 63] for smooth functions, and the starting point in [63] was the expression

$$\tau_y f(x) = V_\kappa^{(x)} \otimes V_\kappa^{(y)} \left[(V_\kappa^{-1} f)(x+y)\right],$$

where V_κ^{-1} denotes the inverse of V_κ, which satisfies $V_\kappa^{-1} f(x) = e^{-\langle y, \mathscr{D}\rangle} f(x)|_{y=0}$. This expression by itself, however, does not provide much useful information on $\tau_y f$. The formula for τ_y for radial functions was proven in [44] under more restrictive conditions, our proof here is taken from an earlier version of [59] and the rest of our development in Section 6.3 follows from the latter paper. Our Section 6.4, convolution and summability also follows the treatment in [59]. Some of the summability method, such as the heat transform and Poisson transforms were studied earlier in [42, 45]. The maximal functions were defined and studied in [59]. Further results on maximal functions were obtained by [1, 17, 18]. Paley–Wiener theorem (Theorem 6.1.10) was proved in [59]. A much more general study of Paley–Wiener theorems for Dunkl transform was given in [35], see also [63].

Many results for the classical Fourier transform can be extended to the Dunkl transforms. For example, in the distributional sense

$$\mathscr{F}_\kappa\left(\frac{P(x)}{\|x\|^{2\lambda_\kappa+2+n-\alpha}}\right) = d_{n,\kappa}^{\alpha} \frac{P(x)}{\|x\|^{n+\alpha}}, \qquad d_{n,\kappa}^{\alpha} = i^{-n} \frac{2^{\lambda_\kappa+1-\alpha}\Gamma(\frac{n+\alpha}{2})}{\Gamma(\lambda_\kappa+1+\frac{n-\alpha}{2})},$$

where $P \in \mathscr{H}_n^d(h_\kappa^2)$ and $0 < \mathrm{Re}\{\alpha\} < 2\lambda_\kappa + 2$, which allows one to define analogues of the Riesz potentials and Bessel potentials for the Dunkl transforms and to study their boundedness in L^p spaces [60]. For the latter purpose, however, we need the boundedness of the translation operator, which however is known, as shown in Section 6.3, to hold, only for $G = \mathbb{Z}_2^d$ or radial functions. Thus, the boundedness can be established at this point only for $G = \mathbb{Z}_2^d$; see [32, 59].

The simplest non-trivial case of the Dunkl transform is the one on the real line associated with the weight function $|x|^\kappa$ that is invariant under $\mathbb{Z}_2$. If f is even, then it agrees with the Hankel transform. For general functions, the difference part comes in and needs to be dealt with. Nevertheless, the structure is relatively simple and many tools in harmonic analysis are accessible. There are numerous papers on Dunkl transforms on the real line. Interested readers should check MathSciNet or arXiv.

Chapter 7

Multiplier Theorems for the Dunkl Transform

For a family of weight functions invariant under a finite reflection group, we prove a transference theorem between the L^p multiplier of h-harmonic expansions on $\mathbb{S}^d$ and that of the Dunkl transform. This theorem is stated together with some related definitions and notations in Section 7.1. The proof of this transference theorem is, however, rather long, so we split it into three parts, which are given in the Sections 7.2, 7.3, and 7.4, respectively. The transference theorem allows us to deduce several useful results for the Dunkl transform on $\mathbb{R}^d$ from the corresponding results for the h-harmonic expansions on $\mathbb{S}^d$. This is done in the last two sections, 7.5 and 7.6. More precisely, in Section 7.5, the transference theorem combined with Theorem 4.4.2, Theorem 4.5.2 is used to establish a Hörmander type multiplier theorem and the Littlewood–Paley inequality for the Dunkl transform on $\mathbb{R}^d$. In Section 7.5, we apply the transference theorem and Theorem 3.3.6 to deduce the convergence of the Bochner–Riesz means of order above the critical index in the weighted L^p spaces for the group $G = \mathbb{Z}_2^d$.

7.1 Introduction

Let h_κ denote the weight function on $\mathbb{R}^d$ defined by (2.1.2), invariant under a finite reflection group G generated by a reduced root system R in $\mathbb{R}^d$. Throughout this chapter, the root system R is normalized so that $\langle \alpha, \alpha \rangle = 2$ for all $\alpha \in R$, and κ denotes a nonnegative multiplicative function on R. For each $g \in G$, we denote by g' the reflection on $\mathbb{R}^{d+1}$ given by

$$x'g' = (xg, x_{d+1}) \text{ for } x' = (x, x_{d+1}) \text{ with } x \in \mathbb{R}^d \text{ and } x_{d+1} \in \mathbb{R}.$$

Then $G' := \{g' : g \in G\}$ is a finite reflection group on $\mathbb{R}^{d+1}$ with a reduced root system $R' := \{(\alpha, 0) : \alpha \in R\}$. Let κ' denote the nonnegative multiplicity function on R' given by $\kappa'(\alpha, 0) = \kappa_\alpha$ for $\alpha \in R$. We denote by $V_{\kappa'}$ the intertwining operator on $C(\mathbb{R}^{d+1})$

associated with the reflection group G' and the multiplicity function κ'. Define

$$h_{\kappa'}(x,x_{d+1}) := h_\kappa(x) = \prod_{\alpha\in R_+} |\langle x,\alpha\rangle|^{\kappa_\alpha}, \quad x\in\mathbb{R}^d,\ x_{d+1}\in\mathbb{R},$$

where R_+ is an arbitrary but fixed positive subsystem of R. Recall that $\mathscr{H}_n^{d+1}(h_{\kappa'}^2)$ denotes the space of h-harmonics of degree n on the sphere $\mathbb{S}^d$, and $\operatorname{proj}_n^{\kappa'} : L^2(\mathbb{S}^d; h_{\kappa'}^2) \to \mathscr{H}_n^{d+1}(h_{\kappa'}^2)$ denotes the orthogonal projection onto the space $\mathscr{H}_n^{d+1}(h_{\kappa'}^2)$. We will keep the notations G', κ', $h_{\kappa'}$ and $V_{\kappa'}$ throughout this chapter. Also, recall that for a given $1\le p\le\infty$, $L^p(\mathbb{R}^d; h_\kappa^2)$ denotes the weighted Lebesgue space on $\mathbb{R}^d$ endowed with the norm

$$\|f\|_{\kappa,p} := \left(\int_{\mathbb{R}^d} |f(y)|^p h_\kappa^2(y)\,dy\right)^{\frac{1}{p}},$$

with the usual change when $p=\infty$.

One of the main goals in this chapter is to show the following transference theorem between the L^p multiplier of h-harmonic expansions on $\mathbb{S}^d$ and that of the Dunkl transform.

Theorem 7.1.1. *Let $m:\ [0,\infty)\to\mathbb{R}$ be a continuous and bounded function, and let U_ε, $\varepsilon>0$, be a family of multiplier operators on $L^2(\mathbb{S}^d; h_{\kappa'}^2)$ given by*

$$\operatorname{proj}_n^{\kappa'}(U_\varepsilon f) = m(\varepsilon n)\operatorname{proj}_n^{\kappa'} f, \quad n=0,1,\dots. \tag{7.1.1}$$

Assume that

$$\sup_{\varepsilon>0} \|U_\varepsilon f\|_{L^p(\mathbb{S}^d;h_{\kappa'}^2)} \le A\|f\|_{L^p(\mathbb{S}^d;h_{\kappa'}^2)}, \quad \forall f\in C(\mathbb{S}^d) \tag{7.1.2}$$

for some $1\le p\le\infty$. Then the function $m(\|\cdot\|)$ defines an $L^p(\mathbb{R}^d;h_\kappa^2)$ multiplier; that is,

$$\|T_m f\|_{L^p(\mathbb{R}^d;h_\kappa^2)} \le c_{d,\kappa} A\|f\|_{L^p(\mathbb{R}^d;h_\kappa^2)}, \quad \forall f\in\mathscr{S}(\mathbb{R}^d),$$

where T_m is an operator initially defined on $L^2(\mathbb{R}^d;h_\kappa^2)$ by

$$\mathscr{F}_\kappa(T_m f)(\xi) = m(\|\xi\|)\mathscr{F}_\kappa f(\xi), \quad f\in L^2(\mathbb{R}^d;h_\kappa^2),\ \xi\in\mathbb{R}^d. \tag{7.1.3}$$

The proof of Theorem 7.1.1 is rather long, so we break it into three parts, given in Sections 7.2, 7.3 and 7.4, respectively. The first part contains several technical lemmas that are crucial for the proof. The second part proves the conclusion under the additional assumption that $|m(t)|\le c_1 e^{-c_2 t}$ for all $t>0$ and some $c_1,c_2>0$. The third part shows how the additional decaying condition on m can be relaxed to yield the desired conclusion.

7.2 Proof of Theorem 7.1.1: part I

A series of technical lemmas used in the proof of Theorem 7.1.1 are proved below. The first lemma reveals a connection between the Dunkl intertwining operators V_κ and $V_{\kappa'}$.

Lemma 7.2.1. *If $f \in \Pi^{d+1}$, then for any $x \in \mathbb{R}^d$ and $x_{d+1} \in \mathbb{R}$,*

$$V_{\kappa'} f(x, x_{d+1}) = V_\kappa[f(\cdot, x_{d+1})](x) = \int_{\mathbb{R}^d} f(\xi, x_{d+1})\, d\mu_x^\kappa(\xi), \tag{7.2.1}$$

where $d\mu_x^\kappa$ is the Borel measure in the integral representation of V_κ in Theorem 2.3.4.

Proof. Clearly, the second equality in (7.2.1) follows directly from Theorem 2.3.4. To show the first equality, we set $V_{\kappa,1} f(x, x_{d+1}) = V_\kappa[f(\cdot, x_{d+1})](x)$ for $f \in C(\mathbb{R}^{d+1})$ and $x \in \mathbb{R}^d$. Since $V_{\kappa'}$ is a linear operator uniquely determined by (2.3.1), it suffices to show that the following conditions are satisfied:

$$V_{\kappa,1}(\mathscr{P}_n^{d+1}) \subset \mathscr{P}_n^{d+1}, \ V_{\kappa,1}(1) = 1, \ \text{and} \ \mathscr{D}_{\kappa',i} V_{\kappa,1} = V_{\kappa,1} \partial_i, \ 1 \le i \le d+1,$$

where we used the notation $\mathscr{D}_{\kappa,i}$ rather than $\mathscr{D}_i$ to denote the Dunkl operators introduced in Definition 2.2.1 to emphasize their dependence on the multiplicative function κ. Indeed, these conditions can be easily verified using the properties of V_κ in (2.3.1), and the following identities, which follow directly from (2.2.1):

$$\mathscr{D}_{\kappa',i} g(x, x_{d+1}) = \mathscr{D}_{\kappa,i}\Big[g(\cdot, x_{d+1})\Big](x), \ \ 1 \le i \le d,$$
$$\mathscr{D}_{\kappa',d+1} g(x, x_{d+1}) = \partial_{d+1} g(x, x_{d+1}), \ \ \text{for } g \in \Pi^{d+1}, x \in \mathbb{R}^d, \text{ and } x_{d+1} \in \mathbb{R}.$$

This completes the proof of Lemma 7.2.1. □

To formulate the next lemma, we define the mapping $\psi : \mathbb{R}^d \to \mathbb{S}^d$ by

$$\psi(x) := (\xi \sin\|x\|, \cos\|x\|) \quad \text{for } x = \|x\|\xi \in \mathbb{R}^d \text{ and } \xi \in \mathbb{S}^{d-1}.$$

Given $N \ge 1$, we denote by $N\mathbb{S}^d := \{x \in \mathbb{R}^{d+1} : \ \|x\| = N\}$ the sphere of radius N in $\mathbb{R}^{d+1}$, and define the mapping $\psi_N : \mathbb{R}^d \to N\mathbb{S}^d$ by

$$\psi_N(x) := N\psi\left(\frac{x}{N}\right) = \left(N\xi \sin\frac{\|x\|}{N}, N\cos\frac{\|x\|}{N}\right) \tag{7.2.2}$$

with $x = \|x\|\xi \in \mathbb{R}^d$ and $\xi \in \mathbb{S}^{d-1}$.

Lemma 7.2.2. *If $f \colon N\mathbb{S}^d \to \mathbb{R}$ is supported in the set $\{x \in N\mathbb{S}^d : \arccos(N^{-1}x_{d+1}) \le 1\}$, then*

$$\int_{\mathbb{S}^d} f(Nx) h_{\kappa'}^2(x)\, d\sigma(x) = N^{-2\lambda_\kappa - 2} \int_{B(0,N)} f(\psi_N(x)) h_\kappa^2(x) \left(\frac{\sin(\|x\|/N)}{\|x\|/N}\right)^{2\lambda_\kappa + 1} dx,$$

where $B(0,N) = \{y \in \mathbb{R}^d : \ \|y\| \le N\}$, and $\lambda_\kappa = \frac{d-2}{2} + |\kappa|$.

Proof. First, using the polar coordinates transformation

$$(\xi, \theta) \in \mathbb{S}^{d-1} \times [0, \pi] \to x := (\xi \sin\theta, \cos\theta) \in \mathbb{S}^d,$$

and the fact that $d\sigma(x) = \sin^{d-1}\theta\, d\theta d\sigma(\xi)$, we obtain

$$\begin{aligned}
&\int_{\mathbb{S}^d} f(Nx)h_{\kappa'}^2(x)\,d\sigma(x)\\
&\quad = \int_0^{\pi}\Big[\int_{\mathbb{S}^{d-1}} f(N\xi\sin\theta, N\cos\theta)h_{\kappa'}^2(\xi\sin\theta,\cos\theta)\,d\sigma(\xi)\Big](\sin\theta)^{d-1}\,d\theta\\
&\quad = \int_0^{1}\Big[\int_{\mathbb{S}^{d-1}} f(N\xi\sin\theta, N\cos\theta)h_{\kappa}^2(\theta\xi)\,d\sigma(\xi)\Big]\left(\frac{\sin\theta}{\theta}\right)^{d-1+2\gamma_\kappa}\theta^{d-1}\,d\theta,
\end{aligned}$$

where the last step uses the identity $h_{\kappa'}(y, y_{d+1}) = h_\kappa(y)$, the fact that h_κ^2 is a homogeneous function of degree $2\gamma_\kappa$, and the assumption that f is supported in the set $\{x \in N\mathbb{S}^d : \arccos(N^{-1}x_{d+1}) \le 1\}$. Using the usual spherical coordinates transformation in $\mathbb{R}^d$, the last double integral equals

$$\begin{aligned}
&\int_{\|y\|\le 1} f\left(\frac{Ny\sin\|y\|}{\|y\|}, N\cos\|y\|\right)h_\kappa^2(y)\left(\frac{\sin\|y\|}{\|y\|}\right)^{2\lambda_\kappa+1}dy\\
&= N^{-d-2\gamma_\kappa}\int_{\|x\|\le N} f\left(N\frac{x}{\|x\|}\sin\frac{\|x\|}{N}, N\cos\frac{\|x\|}{N}\right)h_\kappa^2(x)\left(\frac{\sin(\|x\|/N)}{\|x\|/N}\right)^{2\lambda_\kappa+1}dx\\
&= N^{-2\lambda_\kappa-2}\int_{B(0,N)} f(\psi_N x)h_\kappa^2(x)\left(\frac{\sin(\|x\|/N)}{\|x\|/N}\right)^{2\lambda_\kappa+1}dx,
\end{aligned}$$

where the first step uses the homogeneity of the weight h_κ and the change of variables $y = x/N$. This proves the desired formula. □

Remark 7.2.3. It is easily seen that the restriction $\psi_N\big|_{B(0,N)}$ of the mapping ψ_N on $B(0,N)$ is a bijection from $B(0,N)$ to $\{x \in N\mathbb{S}^d : \arccos(N^{-1}x_{d+1}) \le 1\}$. Thus, given a function $f : B(0,N) \to \mathbb{R}$, there exists a unique function f_N supported in $\{x \in N\mathbb{S}^d : \arccos(N^{-1}x_{d+1}) \le 1\}$ such that

$$f_N(\psi_N x) = f(x), \quad \forall x \in B(0,N). \tag{7.2.3}$$

On the other hand, using Lemma 7.2.2, we have

$$\int_{\mathbb{S}^d} f_N(Nx)h_{\kappa'}^2(x)\,d\sigma(x) = N^{-2\lambda_\kappa-2}\int_{B(0,N)} f(x)h_\kappa^2(x)\left(\frac{\sin(\|x\|/N)}{\|x\|/N}\right)^{2\lambda_\kappa+1}dx. \tag{7.2.4}$$

The formula (7.2.4) will play an important role in our proof of Theorem 7.1.1.

The third lemma asserts that the conclusion of Theorem 6.3.2 holds under a slightly weaker condition.

Lemma 7.2.4. *If $f(x) = f_0(\|x\|)$ is a continuous radial function in $L^2(\mathbb{R}^d; h_\kappa^2)$, then for almost every $y \in \mathbb{R}^d$ and almost every $x \in \mathbb{R}^d$,*

$$\tau_y f(x) = V_\kappa\Big[f_0\Big(\sqrt{\|x\|^2+\|y\|^2-2\|y\|\langle x,\cdot\rangle}\Big)\Big]\Big(\frac{y}{\|y\|}\Big). \tag{7.2.5}$$

Proof. We first choose a sequence of even, C^∞ functions g_j on $\mathbb{R}$ satisfying

$$\sup_{|t|\le 2^{j+1}} |g_j(t)-f_0(t)| \le 2^{-j}\left(\int_0^{2^j} s^{2\lambda_\kappa+1}\,ds\right)^{-\frac{1}{2}}.$$

Let φ_j be an even, C^∞ function on $\mathbb{R}$ such that $\chi_{[2^{-j},2^j]}(|t|) \le \varphi_j(t) \le \chi_{[2^{-j-1},2^{j+1}]}(|t|)$, and let $f_j(x) \equiv f_{j,0}(\|x\|) := g_j(\|x\|)\varphi_j(\|x\|)$ for $x\in\mathbb{R}^d$. Then it is easily seen that $\{f_j\}$ is a sequence of radial Schwartz functions on $\mathbb{R}^d$ satisfying

$$\lim_{j\to\infty}\sup_{2^{-j}\le|t|\le 2^j} |f_{j,0}(t)-f_0(t)| = 0 \tag{7.2.6}$$

and

$$\lim_{j\to\infty}\|f_j-f\|_{L^2(\mathbb{R}^d;h_\kappa^2)} = 0. \tag{7.2.7}$$

Since each f_j is a radial Schwartz function, by Theorem 6.3.2, we obtain

$$\tau_y(f_j)(x) = \int_{\|\xi\|\le 1} f_{j,0}\left(\sqrt{\|x\|^2+\|y\|^2-2\|y\|\langle x,\xi\rangle}\right)d\mu^\kappa_{y/\|y\|}(\xi). \tag{7.2.8}$$

Next, we fix $y\in\mathbb{R}^d$, and set

$$A_n \equiv A_n(y) := \{x\in\mathbb{R}^d:\ 2^{-n}\le \Big|\|x\|-\|y\|\Big| \le \|x\|+\|y\|\le 2^n\},$$

for $n\in\mathbb{N}$ and $n\ge n_0(y) := [\log\|y\|/\log 2]+1$. Since

$$(\|x\|-\|y\|)^2 \le \|x\|^2+\|y\|^2-2\|y\|\,|\langle x,\xi\rangle| \le (\|x\|+\|y\|)^2$$

for all $\|\xi\|\le 1$, it follows by (7.2.6) that

$$\lim_{j\to\infty} f_{j,0}\left(\sqrt{\|x\|^2+\|y\|^2-2\|y\|\langle x,\xi\rangle}\right) = f_0\left(\sqrt{\|x\|^2+\|y\|^2-2\|y\|\langle x,\xi\rangle}\right)$$

uniformly for $x\in A_n(y)$ and $\|\xi\|\le 1$. This, together with (7.2.8) and Theorem 2.3.4 implies that

$$\begin{aligned}\lim_{j\to\infty}\tau_y(f_j)(x) =& \int_{\|\xi\|\le 1} f_0\left(\sqrt{\|x\|^2+\|y\|^2-2\|y\|\langle x,\xi\rangle}\right)d\mu^\kappa_{y/\|y\|}(\xi)\\ =&V_\kappa\left[f_0\left(\sqrt{\|x\|^2+\|y\|^2-2\|y\|\langle x,\cdot\rangle}\right)\right]\left(\frac{y}{\|y\|}\right)\end{aligned}$$

for every $x\in A_n(y)\setminus\{0\}$ and $n\ge n_0(y)$. On the other hand, however, by (7.2.7), we have

$$\lim_{j\to\infty}\|\tau_y(f_j)-\tau_y f\|_{\kappa,2} = 0$$

for all $y \in \mathbb{R}^d$. Thus,

$$\tau_y(f)(x) = V_\kappa\Big[f_0\Big(\sqrt{\|x\|^2+\|y\|^2-2\|y\|\langle x,\cdot\rangle}\Big)\Big]\Big(\frac{y}{\|y\|}\Big),$$

for almost every $x \in A_n(y)$ and all $n \geq n_0(y)$. Finally, observing that the set

$$\mathbb{R}^d \setminus \Big(\bigcup_{n=n_0(y)}^{\infty} A_n(y)\Big) = \{x \in \mathbb{R}^d : \|x\| = \|y\|\}$$

has measure zero in $\mathbb{R}^d$, we deduce the desired conclusion. □

Remark 7.2.5. By Theorem 2.3.4 and the support condition on the measure $d\mu_x^\kappa$,

$$V_\kappa F(rx) = \int_{\mathbb{R}^d} F(r\xi)\, d\mu_x^\kappa(\xi), \quad \text{for all } F \in C(\mathbb{R}^d),\, x \in \mathbb{R}^d, \text{ and } r > 0. \tag{7.2.9}$$

Thus, (7.2.5) can be rewritten more symmetrically as

$$\tau_y f(x) = V_\kappa\Big[f_0\Big(\sqrt{\|x\|^2+\|y\|^2-2\langle x,\cdot\rangle}\Big)\Big](y). \tag{7.2.10}$$

Lemma 7.2.6. *Let $\Phi \in L^1(\mathbb{R}, |x|^{2\lambda_\kappa+1})$ be an even, bounded function on $\mathbb{R}$, and let T_Φ be the operator $L^2(\mathbb{R}^d; h_\kappa^2) \to L^2(\mathbb{R}^d; h_\kappa^2)$ defined by*

$$\mathscr{F}_\kappa(T_\Phi f)(\xi) := \mathscr{F}_\kappa f(\xi)\Phi(\|\xi\|), \quad f \in L^2(\mathbb{R}^d; h_\kappa^2).$$

Then T_Φ has an integral representation

$$T_\Phi f(x) = \int_{\mathbb{R}^d} f(y) K(x,y) h_\kappa^2(y)\, dy,$$

valid for $f \in \mathscr{S}(\mathbb{R}^d)$ and almost every $x \in \mathbb{R}^d$, where

$$K(x,y) = c\int_0^\infty \Phi(s) V_\kappa\Big[\frac{J_{\lambda_\kappa}\big(s\sqrt{\|x\|^2+\|y\|^2-2\langle x,\cdot\rangle}\big)}{\big(s\sqrt{\|x\|^2+\|y\|^2-2\langle x,\cdot\rangle}\big)^{\lambda_\kappa}}\Big](y) s^{2\lambda_\kappa+1}\, ds. \tag{7.2.11}$$

Furthermore, $K(x,y) = K(y,x)$ for almost every $x \in \mathbb{R}^d$ and almost every $y \in \mathbb{R}^d$.

Proof. Let $g(x) = H_{\lambda_\kappa}\Phi(\|x\|)$, where $x \in \mathbb{R}^d$ and H_α denotes the Hankel transform. Since Φ is an even function in $L^1(\mathbb{R}, |x|^{2\lambda_\kappa+1}) \cap L^\infty(\mathbb{R})$, it follows by the properties of the Hankel transform that g is a continuous radial function in $L^2(\mathbb{R}^d; h_\kappa^2)$ and $\mathscr{F}_\kappa g(\xi) = \Phi(\|\xi\|)$. Thus, using (6.4.1), we have

$$T_\Phi f(x) = f *_\kappa g(x) = \int_{\mathbb{R}^d} f(y)\tau_y g(x) h_\kappa^2(y)\, dy$$

for $f \in L^2(\mathbb{R}^d; h_\kappa^2)$. Since g is a continuous radial function in $L^2(\mathbb{R}^d; h_\kappa^2)$, by Lemma 7.2.4 and Remark 7.2.5 it follows that

$$\begin{aligned} K(x,y) := \tau_y g(x) &= V_\kappa\Big[H_{\lambda_\kappa}\Phi\Big(\sqrt{\|x\|^2+\|y\|^2+2\langle x,\cdot\rangle}\Big)\Big](y) \\ &= c\int_0^\infty \Phi(s) V_\kappa\Bigg[\frac{J_{\lambda_\kappa}\big(s\sqrt{\|x\|^2+\|y\|^2-2\langle x,\cdot\rangle}\big)}{\big(s\sqrt{\|x\|^2+\|y\|^2-2\langle x,\cdot\rangle}\big)^{\lambda_\kappa}}\Bigg](y)s^{2\lambda_\kappa+1}\,ds, \end{aligned}$$

where the last step uses (2.3.4), the inequality

$$\left|\Phi(s)\frac{J_{\lambda_\kappa}(rs)}{(rs)^{\lambda_\kappa}}\right| \le c|\Phi(s)|$$

and Fubini's theorem. This proves (7.2.11). The equality $K(x,y) = K(y,x)$ follows from the fact that $\tau_x g(y) = \tau_y g(x)$. □

Our final lemma is a well-known result for ultraspherical polynomials (see, for instance, [53, (8.1.1), p.192]):

Lemma 7.2.7. *For $z \in \mathbb{C}$ and $\mu \ge 0$,*

$$\lim_{k\to\infty} k^{1-2\mu} C_k^\mu\left(\cos\frac{z}{k}\right) = \frac{\Gamma(\mu+\frac{1}{2})}{\Gamma(2\mu)}\left(\frac{z}{2}\right)^{-\mu+\frac{1}{2}} J_{\mu-\frac{1}{2}}(z). \tag{7.2.12}$$

This formula holds uniformly in every bounded region of the complex z-plane.

7.3 Proof of Theorem 7.1.1: part II

In this section, we shall prove Theorem 7.1.1 under the additional assumption that $|m(t)| \le c_1 e^{-c_2 t}$ for all $t > 0$ and some $c_1, c_2 > 0$. By Lemma 7.2.6, the operator T_m has the integral representation

$$T_m f(x) = \int_{\mathbb{R}^d} f(y)K(x,y)h_\kappa^2(y)\,dy,$$

where $K(x,y)$ is given by (7.2.11) with $\Phi = m$. Thus, it is sufficient to prove that

$$I := \left|\int_{\mathbb{R}^d}\int_{\mathbb{R}^d} f(y)g(x)K(x,y)h_\kappa^2(x)h_\kappa^2(y)\,dxdy\right| \le cA \tag{7.3.1}$$

whenever $f \in L^p(\mathbb{R}^d; h_\kappa^2)$ and $g \in L^{p'}(\mathbb{R}^d; h_\kappa^2)$ both have compact supports and satisfy $\|f\|_{L^p(\mathbb{R}^d;h_\kappa^2)} = \|g\|_{L^{p'}(\mathbb{R}^d;h_\kappa^2)} = 1$. Here and in what follows, $\frac{1}{p} + \frac{1}{p'} = 1$.

To this end, we choose a sufficiently large positive number N so that the supports of f and g are both contained in the ball $B(0,N)$. By Remark 7.2.3, there exist functions f_N and g_N, both supported in $\{x \in N\mathbb{S}^d :\ \arccos(N^{-1}x_{d+1}) \le 1\}$ and satisfying

$$f_N\big(\psi_N(x)\big) = f(x),\ \ g_N\big(\psi_N(x)\big) = g(x),\ \ x \in \mathbb{R}^d, \tag{7.3.2}$$

where ψ_N is defined by (7.2.2). It is easily seen from (7.2.4) that

$$\|f_N(N\cdot)\|_{L^p(\mathbb{S}^d;h^2_{\kappa'})} \le cN^{-\frac{2\lambda_\kappa+2}{p}}, \qquad \|g_N(N\cdot)\|_{L^{p'}(\mathbb{S}^d;h^2_{\kappa'})} \le cN^{-\frac{2\lambda_\kappa+2}{p'}}.$$

Thus, using (3.2.1), (3.2.3), (7.1.1), and the assumption (7.1.2) with $\varepsilon = \frac{1}{N}$, we obtain

$$\begin{aligned} I_N &:= N^{2\lambda_\kappa+2} \\ &\quad \times \left|\int_{\mathbb{S}^d}\left[\int_{\mathbb{S}^d}\left(\sum_{n=0}^{\infty} m(N^{-1}n)P_n^{\kappa'}(x,y)\right) f_N(Ny)g_N(Nx)h^2_{\kappa'}(x)h^2_{\kappa'}(y)\,d\sigma(x)\right)d\sigma(y)\right| \\ &\le cA, \end{aligned} \tag{7.3.3}$$

where $P_n^{\kappa'}(x,y) = \frac{n+\lambda_\kappa+1/2}{\lambda_\kappa+1/2}V_{\kappa'}[C_n^{\lambda_\kappa+1/2}(\langle x,\cdot\rangle)](y)$. Setting

$$H_N(x,y) = N^{-2\lambda_\kappa-2}\sum_{n=0}^{\infty} m(N^{-1}n)P_n^{\kappa'}\left(\psi(\frac{x}{N}),\psi(\frac{y}{N})\right),$$

and invoking (7.3.2) and Lemma 7.2.2, we obtain

$$\begin{aligned} I_N = &\left|\int_{\mathbb{R}^d}\left[\int_{\mathbb{R}^d} H_N(x,y)f(y)g(x)h^2_\kappa(x)h^2_\kappa(y)\left(\frac{\sin(\|x\|/N)}{\|x\|/N}\right)^{2\lambda_\kappa+1}\right.\right. \\ &\times \left.\left.\left(\frac{\sin(\|y\|/N)}{\|y\|/N}\right)^{2\lambda_\kappa+1} dx\right]dy\right|. \end{aligned} \tag{7.3.4}$$

On the other hand, setting

$$b_N(\rho,x,y) = N^{-2\lambda_\kappa-2}\sum_{n=0}^{\infty} m(\frac{n}{N})P_n^{\kappa'}\left(\psi(\frac{x}{N}),\psi(\frac{y}{N})\right)\left(\int_{\frac{n}{N}}^{\frac{n+1}{N}} t^{2\lambda_\kappa+1}\,dt\right)^{-1}\chi_{[\frac{n}{N},\frac{n+1}{N})}(\rho),$$

we have

$$H_N(x,y) = \int_0^\infty b_N(\rho,x,y)\rho^{2\lambda_\kappa+1}\,d\rho.$$

Hence, by (7.3.4),

$$\begin{aligned} I_N = &\left|\int_{\mathbb{R}^d}\left[\int_{\mathbb{R}^d}\left(\int_0^\infty b_N(\rho,x,y)\rho^{2\lambda_\kappa+1}\,d\rho\right)f(y)g(x)h^2_\kappa(x)h^2_\kappa(y)\right.\right. \\ &\times \left.\left.\left(\frac{\sin(\|x\|/N)}{\|x\|/N}\right)^{2\lambda_\kappa+1}\left(\frac{\sin(\|y\|/N)}{\|y\|/N}\right)^{2\lambda_\kappa+1} dx\right]dy\right|. \end{aligned} \tag{7.3.5}$$

The key ingredient in our proof is to show that $\lim_{N\to\infty} I_N = cI$, where c is a constant depending only on d and κ. In fact, once this is proven, then the desired estimate (7.3.1) will follow immediately from (7.3.3).

To show $\lim_{N\to\infty} I_N = cI$, we make the following two assertions:

Assertion 1. For any $N > 0$ and $x, y \in \mathbb{R}^d$,

$$|b_N(\rho, x, y)| \le ce^{-c_2\rho},$$

where c is independent of x, y and N.

Assertion 2. For any fixed $x, y \in \mathbb{R}^d$ and $\rho > 0$,

$$\lim_{N\to\infty} b_N(\rho, x, y) = cm(\rho)V_\kappa\Big[\frac{J_{\lambda_\kappa}(\rho u(x,y,\cdot))}{(\rho u(x,y,\cdot))^{\lambda_\kappa}}\Big](y), \tag{7.3.6}$$

where $u(x,y,\xi) = \sqrt{\|x\|^2 + \|y\|^2 - 2\langle x, \xi\rangle}$, and c is a constant depending only on d and κ.

For the moment, we take the above two assertions for granted, and proceed with the proof of Theorem 7.1.1. By Assertion 1 and Hölder's inequality, we can apply the dominated convergence theorem to the integrals in (7.3.5), and obtain

$$\lim_{N\to\infty} I_N = \Big|\int_{\mathbb{R}^d}\Big[\int_{\mathbb{R}^d}\Big(\int_0^\infty \lim_{N\to\infty} b_N(\rho,x,y)\rho^{2\lambda_\kappa+1}\,d\rho\Big)f(y)g(x)h_\kappa^2(x)h_\kappa^2(y)\,dx\Big]dy\Big|,$$

which, using Assertion 2, equals

$$\begin{aligned}
&c\Big|\int_{\mathbb{R}^d}\Big[\int_{\mathbb{R}^d}\Big(\int_0^\infty m(\rho)V_\kappa\Big[\frac{J_{\lambda_\kappa}(\rho u(x,y,\cdot))}{(\rho u(x,y,\cdot))^{\lambda_\kappa}}\Big](y)\rho^{2\lambda_\kappa+1}\,d\rho\Big)f(y)g(x)h_\kappa^2(x)h_\kappa^2(y)dx\Big]dy\Big| \\
&= c\Big|\int_{\mathbb{R}^d}\Big(\int_{\mathbb{R}^d} K(x,y)f(y)g(x)h_\kappa^2(x)h_\kappa^2(y)\,dx\Big)dy\Big| = cI,
\end{aligned}$$

where the second step uses (7.2.11). Thus, we have shown the desired relation $\lim_{N\to\infty} I_N = cI$, assuming Assertions 1 and 2.

Now we return to the proofs of Assertions 1 and 2. We start with Assertion 1. Assume that $\frac{n}{N} \le \rho < \frac{n+1}{N}$ for some $n \in \mathbb{Z}_+$. Then $|m(\frac{n}{N})| \le c_1e^{-c_2\frac{n}{N}} \le ce^{-c_2\rho}$, and $\int_{\frac{n}{N}}^{\frac{n+1}{N}} t^{2\lambda_\kappa+1}\,dt \ge cN^{-1}\rho^{2\lambda_\kappa+1}$. Hence,

$$\begin{aligned}
|b_N(\rho,x,y)| &= N^{-2\lambda_\kappa-2}\Big|m(\frac{n}{N})P_n^{\kappa'}\Big(\psi(\frac{x}{N}),\psi(\frac{y}{N})\Big)\Big|\Big(\int_{\frac{n}{N}}^{\frac{n+1}{N}} t^{2\lambda_\kappa+1}\,dt\Big)^{-1} \\
&\le cN^{-2\lambda_\kappa-1}\rho^{-2\lambda_\kappa-1}e^{-c_2\rho}\frac{n+\lambda_\kappa+1/2}{\lambda_\kappa+1/2}\Big|V_{\kappa'}\Big[C_n^{\lambda_\kappa+1/2}(\langle\psi(\frac{x}{N}),\cdot\rangle)\Big](\psi(\frac{y}{N}))\Big| \\
&\le c(N\rho)^{-2\lambda_\kappa-1}e^{-c_2\rho}n^{2\lambda_\kappa+1} \le ce^{-c_2\rho},
\end{aligned}$$

where we used (3.2.3) in the second step, and the positivity of V_κ and the estimate $|C_n^{\lambda_\kappa+1/2}(t)| \le cn^{2\lambda_\kappa}$ in the third step. This proves Assertion 1.

Next, we show Assertion 2. A straightforward calculation shows that for $\frac{n}{N} \le \rho \le \frac{n+1}{N}$ and $\rho > 0$,

$$\left(\int_{\frac{n}{N}}^{\frac{n+1}{N}} t^{2\lambda_\kappa+1}\,dt\right)^{-1} = \frac{N}{\rho^{2\lambda_\kappa+1}}(1+o_\rho(1)), \quad \text{as } N \to \infty.$$

This implies that for $\frac{n}{N} \le \rho \le \frac{n+1}{N}$ and $\rho > 0$,

$$\begin{aligned} b_N(\rho,x,y) &= m(\rho)\frac{n^{2\lambda_\kappa+1}}{(N\rho)^{2\lambda_\kappa+1}} n^{-2\lambda_\kappa-1} P_n^{\kappa'}\left(\psi\left(\frac{x}{N}\right),\psi\left(\frac{y}{N}\right)\right)(1+o_\rho(1)) \\ &= c m(\rho) n^{-2\lambda_\kappa} V_{\kappa'}\left[C_n^{\lambda_\kappa+1/2}\left(\langle\psi\left(\frac{x}{N}\right),\cdot\rangle\right)\right]\left(\frac{y}{\|y\|}\sin\frac{\|y\|}{N},\cos\frac{\|y\|}{N}\right) + o_\rho(1), \end{aligned}$$

where the continuity of m is used in the first step, and $n^{-2\lambda_\kappa-1}\left|P_n^{\kappa'}\left(\psi(\frac{x}{N}),\psi(\frac{y}{N})\right)\right| \le c$ and $\lim_{N\to\infty}\frac{n^{2\lambda_\kappa+1}}{(N\rho)^{2\lambda_\kappa+1}} = 1$ in the last step. Thus, using Lemma 7.2.1 and (7.2.9), we obtain

$$\begin{aligned} b_N(\rho,x,y) &= c m(\rho) n^{-2\lambda_\kappa}\int_{\mathbb{R}^d} C_n^{\lambda_\kappa+1/2}\left(\frac{1}{\|x\|}\sin\frac{\|x\|}{N}\sum_{j=1}^d x_j\xi_j + \cos\frac{\|y\|}{N}\cos\frac{\|x\|}{N}\right) \\ &\qquad \times d\mu^\kappa_{\frac{y}{\|y\|}\sin\frac{\|y\|}{N}}(\xi) + o_\rho(1) \\ &= c m(\rho) n^{-2\lambda_\kappa}\int_{\|\xi\|\le\|y\|} C_n^{\lambda_\kappa+1/2}(\cos\theta_N(x,y,\xi))\,d\mu_y^\kappa(\xi) + o_\rho(1), \end{aligned} \tag{7.3.7}$$

where $\theta_N(x,y,\xi) \in [0,\pi]$ satisfies

$$\cos\theta_N(x,y,\xi) = \left(\frac{1}{\|x\|\|y\|}\sum_{j=1}^d x_j\xi_j\right)\sin\frac{\|x\|}{N}\sin\frac{\|y\|}{N} + \cos\frac{\|x\|}{N}\cos\frac{\|y\|}{N}.$$

Since

$$\begin{aligned} \cos\theta_N(x,y,\xi) &= 1 - \frac{1}{2N^2}\left(\|x\|^2+\|y\|^2 - 2\sum_{j=1}^d x_j\xi_j\right) + O_{\|x\|,\|y\|}(N^{-4}) \\ &= 1 - \frac{1}{2N^2}u(x,y,\xi)^2 + O_{\|x\|,\|y\|}(N^{-4}), \end{aligned}$$

it follows that

$$\begin{aligned} \theta_N(x,y,\xi) &= 2\arcsin\left(\frac{1}{2N}\sqrt{u(x,y,\xi)^2 + O_{\|x\|,\|y\|}(N^{-2})}\right) \\ &= \frac{1}{N}\sqrt{u(x,y,\xi)^2 + O_{\|x\|,\|y\|}(N^{-2})} + O_{\|x\|,\|y\|}(N^{-2}) \\ &= \frac{\rho u(x,y,\xi) + o_{\|x\|,\|y\|,\rho}(1)}{n}, \end{aligned}$$

where the last step uses the uniform continuity of the function $t \in [0,M] \to \sqrt{t}$ for any $M > 0$, and the relation $\lim_{N\to\infty} \frac{n}{N\rho} = 1$.

Thus, by (7.3.7) and (7.2.12), we have

$$\begin{aligned}
&\lim_{N\to\infty} b_N(\rho,x,y) \\
&= cm(\rho) \lim_{N\to\infty} \int_{\|\xi\|\le\|y\|} n^{-2\lambda_\kappa} C_n^{\lambda_\kappa+1/2}\left(\cos \frac{\rho u(x,y,\xi)+o_{x,y,\rho}(1)}{n}\right) d\mu_y^\kappa(\xi) \\
&= cm(\rho) \int_{\|\xi\|\le\|y\|} (\rho u(x,y,\xi))^{-\lambda_\kappa} J_{\lambda_\kappa}(\rho u(x,y,\xi))\, d\mu_y^\kappa(\xi) \\
&= cm(\rho) V_\kappa\Big[(\rho u(x,y,\cdot))^{-\lambda_\kappa} J_{\lambda_\kappa}(\rho u(x,y,\cdot))\Big](y),
\end{aligned}$$

where we used the fact that $\|C_n^{\lambda_\kappa+1/2}\|_\infty \le cn^{2\lambda_\kappa}$, the bounded convergence theorem, and (7.2.12) in the last step. This proves Assertion 2.

Thus, we have shown the theorem under the additional assumption that $|m(t)| \le c_1 e^{-c_2 t}$.

7.4 Proof of Theorem 7.1.1: part III

In this section, we shall show how to prove Theorem 7.1.1 without the additional assumption that $|m(t)| \le c_1 e^{-c_2 t}$. To this end, let $m_\delta(t) = m(t)e^{-\delta t}$ for $\delta > 0$, and define $T_{m_\delta} : L^2(\mathbb{R}^d, h_\kappa^2) \to L^2(\mathbb{R}^d; h_\kappa^2)$ by

$$\mathscr{F}_\kappa(T_{m_\delta} f)(\xi) = m_\delta(\xi)\mathscr{F}_\kappa f(\xi), \quad f \in L^2(\mathbb{R}^d; h_\kappa^2).$$

By Lemma 3.4.5, for a given $\varepsilon > 0$, $f \mapsto \sum_{n=0}^\infty e^{-n\varepsilon} \operatorname{proj}_n^\kappa f$ is a positive operator on $L^p(\mathbb{S}^d; h_\kappa^2)$ that satisfies

$$\sup_{\varepsilon>0} \Big\| \sum_{n=0}^\infty e^{-n\varepsilon} \operatorname{proj}_n^\kappa f \Big\|_{L^p(\mathbb{S}^d;h_\kappa^2)} \le \|f\|_{L^p(\mathbb{S}^d;h_\kappa^2)}.$$

Thus, applying Theorem 7.1.1 for the already proven case, we have

$$\sup_{\delta>0} \Big\| T_{m_\delta} f \Big\|_{L^p(\mathbb{R}^d;h_\kappa^2)} \le cA\|f\|_{L^p(\mathbb{R}^d;h_\kappa^2)}. \tag{7.4.1}$$

On the other hand, in view of the definition we can decompose the operator T_{m_δ} as

$$T_{m_\delta} f = P_\delta(Tf), \tag{7.4.2}$$

where $\mathscr{F}_\kappa(Tf)(\xi) = m(\|\xi\|)\mathscr{F}_\kappa f(\xi)$ and $\mathscr{F}_\kappa(P_\delta f)(\xi) = e^{-\delta\|\xi\|}\mathscr{F}_\kappa f(\xi)$. The function $P_\delta f$ is called the Poisson integral of f, and it can be expressed as a generalized convolution

$$P_\delta f(x) := (f *_\kappa P_\delta)(x)$$

with

$$P_\delta(x) := 2^{\gamma_\kappa+\frac{d}{2}} \frac{\Gamma(\gamma_\kappa + \frac{d+1}{2})}{\sqrt{\pi}} \frac{\delta}{(\delta^2+\|x\|^2)^{\gamma_\kappa+\frac{d+1}{2}}}.$$

By Lemma 3.4.7, it follows that

$$\lim_{\delta\to 0^+} P_\delta f(x) = f(x), \quad \text{a.e. } x \in \mathbb{R}^d$$

for any $f \in L^q(\mathbb{R}^d; h_\kappa^2)$ with $1 \le q < \infty$. Since m is bounded, $Tf \in L^2(\mathbb{R}^d; h_\kappa^2)$ for $f \in L^2(\mathbb{R}^d; h_\kappa^2)$. Thus, for any $f \in \mathscr{S}$, using (7.4.2),

$$\lim_{\delta\to 0^+} T_{m_\delta} f(x) = \lim_{\delta\to 0^+} P_\delta(Tf)(x) = Tf(x), \ \text{ a.e. } x \in \mathbb{R}^d, \tag{7.4.3}$$

which combined with (7.4.1) and the Fatou theorem implies the desired estimate

$$\|Tf\|_{L^p(\mathbb{R}^d;h_\kappa^2)} \le cA\|f\|_{L^p(\mathbb{R}^d;h_\kappa^2)}.$$

This completes the proof of the theorem. □

7.5 Hörmander's multiplier theorem and the Littlewood–Paley inequality

As a first application of Theorem 7.1.1, we shall prove the following Hörmander type multiplier theorem for the Dunkl transform:

Theorem 7.5.1. *Let $m\colon (0,\infty) \to \mathbb{R}$ be a bounded function satisfying $\|m\|_\infty \le A$ and Hörmander's condition*

$$\frac{1}{R}\int_R^{2R} |m^{(r)}(t)|\,dt \le AR^{-r}, \ \textit{for all } R > 0, \tag{7.5.1}$$

where r is the smallest integer greater than or equal to $\lambda_\kappa + 3/2$. Let T_m be the operator on $L^2(\mathbb{R}^d; h_\kappa^2)$ defined by

$$\mathscr{F}_\kappa(T_m f)(\xi) = m(\|\xi\|)\mathscr{F}_\kappa f(\xi), \ \ \xi \in \mathbb{R}^d.$$

Then

$$\|T_m f\|_{\kappa,p} \le C_p A \|f\|_{\kappa,p}$$

for all $1 < p < \infty$ and $f \in \mathscr{S}(\mathbb{R}^d)$.

Proof. Let $\mu_\ell = m(\ell\varepsilon)$ for $\varepsilon > 0$ and $\ell = 0,1,\ldots$. Then

$$\begin{aligned} |\triangle^r \mu_\ell| &= \varepsilon^r \left| \int_{[0,1]^r} m^{(r)}(\varepsilon t_1 + \cdots + \varepsilon t_r + \varepsilon\ell)\,dt_1\cdots dt_r \right| \\ &\le \int_{[0,\varepsilon]^r} |m^{(r)}(t_1+\cdots+t_r+\varepsilon\ell)|\,dt_1\cdots dt_r \le \varepsilon^{r-1}\int_{\varepsilon\ell}^{\varepsilon(r+\ell)} |m^{(r)}(t)|\,dt. \end{aligned}$$

This implies that, for $2^j \geq r$,

$$2^{j(r-1)} \sum_{l=2^j}^{2^{j+1}} |\Delta^r \mu_l| \leq 2^{j(r-1)} \varepsilon^{r-1} \sum_{l=2^j}^{2^{j+1}} \int_{\varepsilon\ell}^{\varepsilon(r+\ell)} |m^{(r)}(t)|\,dt$$
$$\leq (r-1) 2^{j(r-1)} \varepsilon^{r-1} \int_{2^j\varepsilon}^{\varepsilon(2^{j+1}+r)} |m^{(r)}(t)|\,dt$$
$$\leq 2^{j(r-1)} (r-1) \varepsilon^{r-1} \int_{2^j\varepsilon}^{2^{j+2}\varepsilon} |m^{(r)}(t)|\,dt \leq c_r A,$$

where the last step uses (7.5.1). On the other hand, however, for $2^j \leq r$, we have

$$2^{j(r-1)} \sum_{l=2^j}^{2^{j+1}} |\Delta^r \mu_l| \leq c_r \max_j |\mu_j| \leq c_r A.$$

Thus, using Theorem 4.4.2, we deduce

$$\sup_{\varepsilon>0} \left\| \sum_{n=0}^{\infty} m(\varepsilon n) \operatorname{proj}_n^{\kappa'} f \right\|_{L^p(\mathbb{S}^d; h_{\kappa'}^2)} \leq c \|f\|_{L^p(\mathbb{S}^d; h_{\kappa'}^2)}.$$

The desired conclusion then follows by Theorem 7.1.1. □

Remark 7.5.2. Hörmander's condition is normally stated in the form

$$\left(\frac{1}{R} \int_R^{2R} |m^{(r)}(t)|^2\,dt \right)^{\frac{1}{2}} \leq A R^{-r}, \quad \text{for all } R > 0; \tag{7.5.2}$$

see, for instance, [31, Theorem 5.2.7]. Clearly, the condition (7.5.1) in Theorem 7.5.1 is weaker than (7.5.2). On the other hand, however, Theorem 7.5.1 is applicable only to radial multipliers $m(\|\cdot\|)$.

Corollary 7.5.3. *Let Φ be an even C^∞-function that is supported in the set $\{x \in \mathbb{R} : \frac{9}{10} \leq |x| \leq \frac{21}{10}\}$ and satisfies either*

$$\sum_{j\in\mathbb{Z}} \Phi(2^{-j}\xi) = 1, \quad \xi \in \mathbb{R} \setminus \{0\},$$

or

$$\sum_{j\in\mathbb{Z}} |\Phi(2^{-j}\xi)|^2 = 1, \quad \xi \in \mathbb{R} \setminus \{0\}.$$

Let $\triangle_j$ be an operator defined by

$$\mathscr{F}_\kappa(\triangle_j f)(\xi) = \Phi(2^{-j}\|\xi\|) \mathscr{F}_\kappa f(\xi), \quad \xi \in \mathbb{R}^d.$$

Then we have

$$\|f\|_{\kappa,p} \sim_{\kappa,p} \|(\sum_{j\in\mathbb{Z}} |\triangle_j f|^2)^{\frac{1}{2}}\|_{\kappa,p}$$

for all $f \in L^p(\mathbb{R}^d; h_\kappa^2)$ and $1 < p < \infty$.

Proof. Corollary 7.5.3 follows directly from Theorem 7.5.1. Since the proof runs along the same line as that of Theorem 4.5.2, we omit the details. □

7.6 Convergence of the Bochner–Riesz means

Recall that the Bochner–Riesz means of order $\delta > -1$ for the Dunkl transform $S_R^\delta f(x) \equiv S_R^\delta(h_\kappa^2; f)(x)$ are defined by (6.4.4). According to Theorem 6.4.7, if $\delta > \lambda_\kappa + \frac{1}{2} := \frac{d-1}{2} + \gamma_\kappa$ and $1 \le p \le \infty$, then

$$\sup_{R>0} \|S_R^\delta(h_\kappa^2; f)\|_{\kappa,p} \le c\|f\|_{\kappa,p}. \tag{7.6.1}$$

Our next result concerns the critical indices for the validity of (7.6.1) in the case of $G = \mathbb{Z}_2^d$:

Theorem 7.6.1. *Suppose that* $G = \mathbb{Z}_2^d$, $f \in L^p(\mathbb{R}^d; h_\kappa^2)$, $1 \le p \le \infty$, *and* $|\frac{1}{p} - \frac{1}{2}| \ge \frac{1}{2\lambda_\kappa+3}$. *Then* (7.6.1) *holds if and only if*

$$\delta > \delta_\kappa(p) := \max\Big\{(2\lambda_\kappa + 2)\Big|\frac{1}{p} - \frac{1}{2}\Big| - \frac{1}{2}, 0\Big\}. \tag{7.6.2}$$

Proof. We start with the proof of the sufficiency. Assume that $\kappa := (\kappa_1, \dots, \kappa_d)$ and $h_\kappa(x) := \prod_{j=1}^d |x_j|^{\kappa_j}$. Let $\kappa' = (\kappa, 0)$ and $h_{\kappa'}(x, x_{d+1}) = h_\kappa(x)$ for $x \in \mathbb{R}^d$ and $x_{d+1} \in \mathbb{R}$. Set $m(t) = (1-t^2)_+^\delta$. By the equivalence of the Riesz and the Cesàro summability methods of order $\delta \ge 0$ (see [30]), we deduce from Theorem 3.3.8 that

$$\sup_{\varepsilon>0} \Big\| \sum_{n=0}^\infty m(\varepsilon n) \operatorname{proj}_n^{\kappa'} f \Big\|_{L^p(\mathbb{S}^d; h_{\kappa'}^2)} \le c\|f\|_{L^p(\mathbb{S}^d; h_{\kappa'}^2)}$$

whenever $|\frac{1}{p} - \frac{1}{2}| \ge \frac{1}{2\sigma_{\kappa'}+2}$ and $\delta > \delta_{\kappa'}(p)$, where $\sigma_{\kappa'} = \lambda_\kappa + \frac{1}{2}$ and $\delta_{\kappa'}(p) = \delta_\kappa(p)$. Thus, invoking Theorem 7.1.1, we conclude that for $\delta > \delta_\kappa(p)$,

$$\|S_1^\delta(h_\kappa^2; f)\|_{\kappa,p} \le c\|f\|_{\kappa,p}.$$

The estimate (7.6.1) then follows by dilation. This proves the sufficiency.

The necessity part of the theorem follows from the corresponding result for the Hankel transform. To see this, let $f(x) = f_0(\|x\|)$ be a radial function in $L^p(\mathbb{R}^d, h_\kappa^2)$. Using (6.4.4) and Theorem 6.2.11 (vii), we have

$$S_R^\delta(h_\kappa^2; f)(x) = \int_0^R \Big(1 - \frac{r^2}{R^2}\Big)^\delta H_{\lambda_\kappa} f_0(r) r^{2\lambda_\kappa+1} \Big[\int_{\mathbb{S}^{d-1}} E_\kappa(ix, ry') h_\kappa^2(y')\, d\sigma(y')\Big] dr.$$

However, by [60, Proposition 2.3] applied to $n = 0$ and $g = 1$, we have

$$\int_{\mathbb{S}^{d-1}} E_\kappa(ix, ry') h_\kappa^2(y')\, d\sigma(y') = c\Big(\frac{r\|x\|}{2}\Big)^{-\lambda_\kappa} J_{\lambda_\kappa}(r\|x\|).$$

It follows that

$$\begin{aligned} S_R^\delta(h_\kappa^2; f)(x) &= c\int_0^R \Big(1 - \frac{r^2}{R^2}\Big)^\delta H_{\lambda_\kappa} f_0(r) \Big(\frac{r\|x\|}{2}\Big)^{-\lambda_\kappa} J_{\lambda_\kappa}(r\|x\|) r^{2\lambda_\kappa+1}\, dr \\ &= c\widetilde{S}_R^\delta f_0(\|x\|), \end{aligned}$$

where $\widetilde{S}_R^\delta$ denotes the Bockner–Riesz mean of order δ for the Hankel transform H_{λ_κ}. However, it is known (see [66]) that $\widetilde{S}_R^\delta$, $0 < \delta < \lambda_\kappa + \frac{1}{2}$, is bounded on $L^p((0,\infty), t^{2\lambda_\kappa+1})$ if and only if

$$\frac{2\lambda_\kappa + 2}{\lambda_\kappa + \delta + 3/2} < p < \frac{2\lambda_\kappa + 2}{\lambda_\kappa - \delta + 1/2}. \tag{7.6.3}$$

Thus, to complete the Proof of the necessity part of the theorem, by (7.6.3), we just need to observe that if $f(x) = f_0(\|x\|)$ is a radial function in $L^p(\mathbb{R}^d; h_\kappa^2)$, Then

$$\|f\|_{\kappa,p} = c\|f_0\|_{L^p(\mathbb{R};|x|^{2\lambda_\kappa+1})}. \qquad \square$$

7.7 Notes and further results

Most of the results in this chapter were proved in [10]. In the case of ordinary Fourier transform and spherical harmonics, Theorem 7.1.1 is due to Bonami and Clerc [3, Theorem 1.1].

In the unweighted case, for the classical Fourier transform, Theorem 7.6.1 is well known, and in fact, it is a consequence of the following Tomas–Stein restriction theorem (see, for instance, [31, Section 10.4]):

$$\|\widehat{f}\|_{L^2(\mathbb{S}^{d-1})} \le c_p \|f\|_{L^p(\mathbb{R}^d)}, \quad 1 \le p \le \frac{2d+2}{d+3}, \tag{7.7.1}$$

where $\widehat{f}$ denotes the usual Fourier transform of f. In the weighted case, while estimates similar to (7.7.1) can be proved for the Dunkl transform $\mathscr{F}_\kappa f$ (see [38, Theorem 4.1]), they do not seem to be enough for the proof of Theorem 7.6.1. A similar fact was indicated in [14] for the case of the Cesàro means for h-harmonic expansions on the unit sphere, where global estimates for the projection operators have to be replaced with more delicate local estimates, which are significantly more difficult to prove.

Bibliography

[1] C. Abdelkefi, Dunkl operators on $\mathbb{R}^d$ and uncentered maximal function, *J. Lie Theory* **20** (2010), 113–125.

[2] R. Askey, R.G. Andrews and R. Roy, *Special Functions*, Encyclopedia of Mathematics and its Applications **71**, Cambridge University Press, Cambridge, 1999.

[3] A. Bonami and J-L. Clerc, Sommes de Cesàro et multiplicateurs des développements en harmoniques sphériques, *Trans. Amer. Math. Soc.* **183** (1973), 223–263.

[4] R.R. Coifman and G. Weiss, *Analyse Harmonique Non-commutative sur certain Espaces Homogenes*, Lecture Notes in Math. **242**, Springer, Berlin, 1972.

[5] L. Colzani, M.H. Taibleson and G. Weiss, Maximal estimates for Cesàro and Riesz means on spheres, *Indiana Univ. Math. J.* **33** (1984), 873–889.

[6] F. Dai and Z. Ditzian, Combinations of multivariate averages, *J. Approx. Theory* **131** (2004), 268–283.

[7] F. Dai and Z. Ditzian, Littlewood–Paley theory and sharp Marchaud inequality, *Acta Sci. Math. (Szëged)* **71** (2005), 65–90.

[8] F. Dai and Z. Ditzian, Cesàro summability and Marchaud inequality, *Constr. Approx.* **25** (2007), 73–88.

[9] F. Dai, Z. Ditzian and S. Tikhonov, Sharp Jackson inequalities, *J. Approx. Theory* **151** (2008), no. 1, 86–112.

[10] F. Dai and H.P. Wang, A transference theorem for the Dunkl transform and its applications, *J. Funct. Anal.* **258** (2010), 4052–4074.

[11] F. Dai and H.P. Wang, Optimal quadrature in weighted Besov spaces with A_∞ weights on multivariate domains, *Constr. Approx* **37** (2013), no. 2, 167–194.

[12] F. Dai and Y. Xu, Maximal function and multiplier theorem for weighted space on the unit sphere, *J. Funct. Anal.* **249** (2007), 477–504.

[13] F. Dai and Y. Xu, Cesàro means of orthogonal expansions in several variables, *Constr. Approx.* **29** (2009), 129–155.

[14] F. Dai and Y. Xu, Boundedness of projection operators and Cesàro means in weighted L^p space on the unit sphere. *Trans. Amer. Math. Soc.* **361** (2009), 3189–3221.

[15] F. Dai and Y. Xu, Moduli of smoothness and approximation on the unit sphere and the unit ball, *Advances in Math.* **224** (2010), 1233–1310.

[16] F. Dai and Y. Xu, *Approximation Theory and Harmonic Analysis on Spheres and Balls*, Springer Monographs in Mathematics, Springer, New York, 2013.

[17] L. Deleaval, Two results on the Dunkl maximal operator, *Studia Math.* **203** (2011), 47–68.

[18] L. Deleaval, Fefferman–Stein inequalities for the $\mathbb{Z}_2^d$ Dunkl maximal operator, *J. Math. Anal. Appl.* **360** (2009), 711–726.

[19] R.A. DeVore and G.G. Lorentz, *Constructive Approximation*, Springer, New York, 1993.

[20] Z. Ditzian, On the Marchaud-type inequality, *Proc. Amer. Math. Soc.* **103** (1988) 198–202.

[21] Z. Ditzian, Fractional derivatives and best approximation, *Acta Math. Hungar.* **81** (1998), no. 4, 323–348.

[22] Z. Ditzian and A. Prymak, Sharp Marchaud and converse inequalities in Orlicz spaces, *Proc. Amer. Math. Soc.* **135** (2007), 1115-1121.

[23] Z. Ditzian and A. Prymak, Convexity, moduli of smoothness and a Jackson-type inequality, *Acta Math. Hungar.* **130** (2011), no. 3, 254–285.

[24] Z. Ditzian and S. Tikhonov, Ul'yanov and Nikol'skii-type inequalities, *J. Approx. Theory* **133** (2005), 100–133.

[25] C.F. Dunkl, Differential-difference operators associated to reflection groups, *Trans. Amer. Math. Soc.* **311** (1989), 167–183.

[26] C.F. Dunkl, Integral kernels with reflection group invariance, *Can. J. Math.* **43** (1991), 1213–1227.

[27] C.F. Dunkl, Hankel transforms associated to finite reflection groups, in *Hypergeometric Functions on Domains of Positivity, Jack Polynomials, and Applications* (Tampa, FL, 1991), 123–138, Contemporary Mathematics **138**, American Math. Society, Providence, RI. 1992.

[28] C.F. Dunkl, Intertwining operator associated to the group S_3, *Trans. Amer. Math. Soc.* **347** (1995), 3347–3374.

[29] C.F. Dunkl and Y. Xu, *Orthogonal Polynomials of Several Variables*, second edition, Encyclopedia of Mathematics and its Applications **155**, Cambridge University Press, Cambridge, 2014.

[30] J.J. Gergen, Summability of double Fourier series, *Duke Math. J.* **3** (1937), no. 2, 133–148.

[31] L. Grafakos, *Classical and Modern Fourier Analysis*, Pearson Education, Inc., Upper Saddle River, NJ, 2004.

[32] S. Hassani, S. Mustapha and M. Sifi, Riesz potentials and fractional maximal function for the Dunkl transform, *J. Lie Theory* **19** (2009), 725–734.

[33] T.P. Hytnen, Littlewood–Paley–Stein theory for semigroups in UMD spaces. *Rev. Mat. Iberoam.* **23** (2007), no. 3, 973–1009.

[34] M.F.E. de Jeu, The Dunkl transform, *Invent. Math.* **113** (1993), 147–162.

[35] M.F.E. de Jeu, Paley–Wiener theorems for the Dunkl transform, *Trans. Amer. Math. Soc.* **358** (2006), 4225–4250.

[36] T.H. Koornwinder, A new proof of a Paley–Wiener theorem for the Jacobi transform, *Ark.Mat.* **13** (1975), 145–159.

[37] Zh.K. Li, and Y. Xu, Summability of orthogonal expansions of several variables, *J. Approx. Theory*, **122** (2003), 267–333.

[38] L. M. Liu, *Harmonic analysis associated with finite reflection groups*, PhD thesis, 2004, Capital Normal University, Beijing, China.

[39] H. Mejjaoli and K. Trimèche, On a mean value property associated with the Dunkl Laplacian operator and applications, *Integral Transform. Spec. Funct.* **12** (2001), 279–302.

[40] C. Müller, *Analysis of Spherical Symmetries in Euclidean Spaces*, Springer, New York, 1997.

[41] M. Rösler, Bessel-type signed hypergroups on $\mathbb{R}$, in: H. Heyer, A. Mukherjea (eds.), *Probability Measures on Groups and Related Structures XI*, Proc. Oberwolfach 1994, World Scientific, Singapore, 1995, 292–304.

[42] M. Rösler, Generalized Hermite polynomials and the heat equation for Dunkl operators, *Comm. Math. Phys.* **192** (1998), 519–542.

[43] M. Rösler, Positivity of Dunkl's intertwining operator, *Duke Math. J.* **98** (1999), 445–463.

[44] M. Rösler, A positive radial product formula for the Dunkl kernel, *Trans. Amer. Math. Soc.* **355** (2003), 2413–2438.

[45] M. Rösler and M. Voit, Markov processes associated with Dunkl operators, *Adv. Appl. Math.* **21** (1998), 575–643.

[46] W. Rudin, *Real and Complex Analysis*, 3rd ed., McGraw-Hill, New York, 1987.

[47] B. Simonov and S. Tikhonov, Sharp Ul'yanov-type inequalities using fractional smoothness, *J. Approx. Theory* **162** (2010), 1654–1684.

[48] C.D. Sogge, Oscillatory integrals and spherical harmonics, *Duke Math. J.* **53** (1986), 43–65.

[49] E.M. Stein, *Topics in Harmonic Analysis Related to the Littlewood-Paley Theory*, Annals of Mathematical Studies, Vol. 63. Princeton Univ. Press, Princeton, NJ, 1970.

[50] E.M. Stein, *Singular Integrals and Differentiability Properties of Functions*, Princeton University Press, Princeton, NJ, 1970.

[51] E.M. Stein, *Harmonic Analysis: Real-Variable Methods, Orthogonality, and Oscillatroy Integrals*, Princeton Univ. Press, Princeton, NJ, 1993.

[52] E.M. Stein and G. Weiss, *Introduction to Fourier Analysis on Euclidean Spaces*, Princeton Univ. Press, Princeton, NJ, 1971.

[53] G. Szegö, *Orthogonal Polynomials*, Amer. Math. Soc. Colloq. Publ. Vol. 23, Providence, RI, 4th edition, 1975.

[54] S. Tikhonov and W. Trebels, Ulyanov-type inequalities and generalized Liouville derivatives, *Proc. Roy. Soc. Edinburgh Sect. A* **141** (2011), 205–224.

[55] M.F. Timan, Converse theorems of the constructive theory of functions in the spaces L_p, *Math. Sbornic* **46** (1958), no. 88, 125–132.

[56] M.F. Timan, On Jackson's theorem in L_p spaces, *Ukr. Mat. Zh.* **18** (1966), no. 1, 134–137 (Russian).

[57] A.F. Timan, *Theory of Approximation of Functions of a Real Variable*, Translated from the Russian by J. Berry. Translation edited and with a preface by J. Cossar. Reprint of the 1963 English translation. Dover Publ. Inc., Mineola, New York, 1994.

[58] R.M. Trigub and E.S. Belinsky, *Fourier Analysis and Approximation of Functions*, Kluwer Academic Publisher, 2004.

[59] S. Thangavelu and Y. Xu, Generalized translation and convolution operator for Dunkl transform, *J. d'Analyse Mathematique* **97** (2005), 25–56.

[60] S. Thangavelu and Y. Xu, Riesz transform and singular integrals for Dunkl transform, *J. Comp. Appl. Math.* **199** (2007), 181–195.

[61] V. Totik, Sharp converse theorem of L^p polynomial approximation, *Constr. Approx.* **4** (1988), 419–433.

[62] K. Trimèche, The Dunkl intertwining operator on spaces of functions and distributions and integral representation of its dual. *Integral Transform. Spec. Funct.* **12** (2001), 349–374.

[63] K. Trimèche, Paley–Wiener theorems for the Dunkl transform and Dunkl translation operators, *Integral Transforms and Special Functions*, **13** (2002), 17–38.

[64] K.Y. Wang and L.Q. Li, *Harmonic Analysis and Approximation on the Unit Sphere*. Science Press, Beijing, 2000.

[65] G.N. Watson, *A Treatise on the Theory of Bessel Functions*, 2nd edition, Cambridge University Press, London, 1962.

[66] G.V. Welland, Norm convergence of Riesz–Bochner means for radial functions, *Can. J. Math.* **27**(1975), no. 1, 176–185.

[67] P. Wojtaszczyk, On values of homogeneous polynomials in discrete sets of points, *Studia Math.* **84**(1986), no. 1, 97–104.

[68] Y. Xu, Integration of the intertwining operator for h-harmonic polynomials associated to reflection groups, *Proc. Amer. Math. Soc.* **125** (1979), 2963–2973.

[69] Y. Xu, Orthogonal polynomials for a family of product weight functions on the spheres, *Canad. J. Math.* **49** (1997), 175–192.

[70] Y. Xu, A product formula for Jacobi polynomials, in: *Special Functions*, Proceedings of International Workshop, Hong Kong, June 21-25, 1999, World Sci. Publ., 2000, 423–430.

[71] Y. Xu, Funk–Hecke formula for orthogonal polynomials on spheres and on balls, *Bull. London Math. Soc.* **32** (2000), 447–457.

[72] Y. Xu, Weighted approximation of functions on the unit sphere, *Constructive Approx.* **21** (2005), 1–28.

[73] Y. Xu, Almost everywhere convergence of orthogonal expansions of several variables, *Const. Approx.* **22** (2005), 67–93.

[74] Y. Xu, Generalized translation operator and approximation in several variables, *J. Comp. Appl. Math.* **178** (2005), 489–512.

[75] A. Zygmund, *Trigonometric Series*, Cambridge U. Press, 1959.

Index

Advanced Courses in Mathematics – CRM Barcelona (ACM)

Edited by
Carles Casacuberta, Universitat de Barcelona, Spain

Since 1995 the Centre de Recerca Matemàtica (CRM) has organised a number of Advanced Courses at the post-doctoral or advanced graduate level on forefront research topics in Barcelona. The books in this series contain revised and expanded versions of the material presented by the authors in their lectures.

■ **Böckle, G. / Burns, D. / Goss, D. / Thakur, D. / Trihan, F. / Ulmer, D.**, Arithmetic Geometry over Global Function Fields (2014).
ISBN 978-3-0348-0852-1

This volume collects the texts of five courses given in the Arithmetic Geometry Research Programme 2009–2010 at the CRM Barcelona. All of them deal with characteristic *p* global fields; the common theme around which they are centered is the arithmetic of *L*-functions (and other special functions), investigated in various aspects. Three courses examine some of the most important recent ideas in the positive characteristic theory discovered by Goss (a field in tumultuous development, which is seeing a number of spectacular advances): they cover respectively crystals over function fields (with a number of applications to *L*-functions of *t*-motives), gamma and zeta functions in characteristic *p*, and the binomial theorem. The other two are focused on topics closer to the classical theory of abelian varieties over number fields: they give respectively a thorough introduction to the arithmetic of Jacobians over function fields (including the current status of the BSD conjecture and its geometric analogues, and the construction of Mordell-Weil groups of high rank) and a state of the art survey of Geometric Iwasawa Theory explaining the recent proofs of various versions of the Main Conjecture, in the commutative and non-commutative settings.

■ **Asaoka, M. / El Kacimi Alaoui, A. / Hurder, S. / Richardson, K.**, Foliations: Dynamics, Geometry and Topology (2014).
ISBN 978-3-0348-0870-5

The lectures by A. El Kacimi Alaoui offer an introduction to foliation theory, with emphasis on examples and transverse structures. S. Hurder's lectures apply ideas from smooth dynamical systems to develop useful concepts in the study of foliations, like limit sets and cycles for leaves, leafwise geodesic flow, transverse exponents, stable manifolds, Pesin theory, and hyperbolic, parabolic, and elliptic types of foliations, all of them illustrated with examples. The lectures by M. Asaoka are devoted to the computation of the leafwise cohomology of orbit foliations given by locally free actions of certain Lie groups, and its application to the description of the deformation of those actions. In the lectures by K. Richardson, he studies the geometric and analytic properties of transverse Dirac operators for Riemannian foliations and compact Lie group actions, and explains a recently proved index formula.

■ **Alesker, S. / Fu, J. H. G.**, Integral Geometry and Valuations (2014).
ISBN 978-3-0348-0873-6

The first part of this book, by Semyon Alesker, provides an introduction to the theory of convex valuations with emphasis on recent developments. In particular, it presents the new structures on the space of valuations discovered after Alesker's irreducibility theorem. The newly developed theory of valuations on manifolds is also described.

In the second part, Joseph H. G. Fu gives a modern introduction to integral geometry in the sense of Blaschke and Santaló. The approach is new and based on the notions and tools presented in the first part. This original viewpoint not only enlightens the classical integral geometry of euclidean space, but it also allows the computation of kinematic formulas in other geometries, such as hermitian spaces.

■ **Cruz-Uribe, D. / Fiorenza, A. / Ruzhansky, M. / Wirth, J.**, Variable Lebesgue Spaces and Hyperbolic Systems (2014).
ISBN 978-3-0348-0839-2

■ **Berger, L. / Böckle, G. / Dembélé, L. / Dimitrov, M. / Dokchitser, T. / Voight, J.**, Elliptic Curves, Hilbert Modular Forms and Galois Deformations (2013).
ISBN 978-3-0348-0617-6

■ **Cominetti, R. / Facchinei, F. / Lasserre, J. B.**, Modern Optimization Modelling Techniques (2012).
ISBN 978-3-0348-0290-1

Printing: Ten Brink, Meppel, The Netherlands
Binding: Ten Brink, Meppel, The Netherlands